Applied Computing

Springer

London
Berlin
Heidelberg
New York
Barcelona
Budapest
Hong Kong
Milan
Paris
Santa Clara
Singapore
Tokyo

Applied Computing

Series Editors
Professor Ray Paul and Professor Peter Thomas

The Springer-Verlag Series on Applied Computing is an advanced series of innovative textbooks that span the full range of topics in applied computing technology.

Books in the series provide a grounding in theoretical concepts in computer science alongside real-world examples of how those concepts can be applied in the development of effective computer systems.

The series should be essential reading for advanced undergraduate and postgraduate students in computing and information systems.

Books in the series are contributed by international specialist researchers and educators in applied computing who draw together the full range of issues in their specialist area into one concise authoritative textbook.

Titles already available:

Forthcoming titles include:

Deryn Graham and Anthony Barrett

Knowledge-Based Image Processing Systems

With 36 figures

 Springer

Deryn Graham and Anthony Barrett
Department of Computer Science and Information Systems at St. John's,
Brunel University, Uxbridge, Middlesex UB8 3PH, UK

Series Editors

Professor Peter J. Thomas, BA (Hons), PhD, AIMgt, FRSA, FVRS
Centre for Personal Information Management, University of the West of
England, Coldharbour Lane, Bristol, BS16 1QY, UK

Professor Ray J. Paul, BSc, MSc, PhD
Department of Computer Science and Information Systems at St. John's,
Brunel University, Kingston Lane, Uxbridge, Middlesex UB8 3PH, UK

ISBN 3-540-76027-X Springer-Verlag Berlin Heidelberg New York

British Library Cataloguing in Publication Data
Graham, Deryn
    Knowledge-based image processing systems. - (Applied computing)
    1. Expert systems (Computer science)   2. Image processing
    I. Title   II. Barrett, Anthony N.
    006.4'2
    ISBN 354076027X

Library of Congress Cataloging-in-Publication Data
Graham, Deryn, 1961-
    Knowledge-based image processing systems / Deryn Graham and Anthony Barret.
        p.   cm.
    Includes bibliographical references and index.
    ISBN 3-540-76027-X (pbk. : alk. paper)
    1. Image processing–Data processing.   2. Expert systems (Computer science)
    I. Barrett, Anthony N.   II. Title.
    TA 1637.G72 1996
    006.3'7–dc21   96-45110

Typesetting: EXPO Holdings, Malaysia
Printed and bound at the Athenæum Press Ltd., Gateshead, Tyne and Wear
34/3830-543210 Printed on acid-free paper

Preface

Knowledge-based systems and *Expert systems*, are systems developed as a result of work in Artificial Intelligence. An expert system is a knowledge-based system with an evaluated level of performance, close to that of an expert. Knowledge-based systems and expert systems are often not distinguished i.e. if an expert is involved, then usually the implication is that it is an expert system. An expert system may be defined as a computer program applied to problem solving associated with a significant degree of human expertise. Such a program uses knowledge and some form of inference mechanism to achieve this.

Image Processing can be considered as consisting of two components. The first deals with the application of a range of mathematical transformations to an image such as a satellite picture or an x-radiograph for example stored in a digital form on the computer. The purpose of the application is to improve the quality of the image and where possible, to render the image so that features not visible to the eye in the original become visible within the rendered or transformed image. The second component, deals with the extraction of features from within the image for subsequent analysis. Image Processing currently plays an important part in many scientific, engineering and commercial environments.

Knowledge-based systems or Expert systems have been applied to numerous application domains e.g. medicine. Likewise, Image Processing has also been applied to many domains, however, although both fields may often address common application areas, this has been to a large extent mutually independent. A combined knowledge-based systems and image processing approach is appropriate to many of the problems each field is independently addressing. Few systems currently exhibit a truly combined approach, most likely due to research communities often working in isolation from each other.

In themselves, each field requires several hundred pages to cover sufficient material necessary to provide a definitive text. In the space available therefore, we have been obliged to select topics specific to each area in order that a reader of limited familiarity with either knowledge-based systems or image processing should acquire a working knowledge of both fields.

This book is aimed at final year and postgraduate students with some background in computer science, and researchers requiring a brief introduction to either area: knowledge-based systems or image processing. The book is in three parts, parts 1 and 2 are designed to provide overviews of each of the areas (each part equating to a half semester course), part III gives a report on current work in which the two converge before describing some of the issues involved in designing future knowledge-based image processing systems. Sample questions are also provided for parts 1 and 2 in an appendix (A).

This book attempts to give an insight into the two areas of research and current systems, and suggests a way forward for designing future systems.

Brunel, 1996   Deryn Graham
                                                              Anthony Barrett

Acknowledgements

We wish to acknowledge the following people for their support and assistance in producing this book:

Dr Monica Jordan, Director of the National Institute for Biological Standards and Controls (NIBSC), Potters Bar. Collaborative research with Dr Jordan greatly motivating the production of this book. We are also grateful for permission given to print several diagrams provided by NIBSC.

Miss Jagdip Basra, for her assistance in collecting literature for Part 3 of this book.

Mrs Pam Osborne (nee Menday), for her secretarial efforts in relation to Parts 2 and 3.

Dr George Goodsell,for his help in proof reading Part 1.

Mr Neil Middleton, who read the proofs for chapter 5, and suggested several useful improvements.

Brunel University's Research Committee, for granting a Brunel Research Initiative and Enterprise Fund (BRIEF) award which provided funding for research to which this book is related.

About the Authors

Deryn Graham

Deryn Graham lectures on Artificial Intelligence in the department of Computer Science and Information Systems at Brunel University. Her Ph.D. is in the area of applied knowledge-based systems. She currently holds a Brunel Research Initiative and Enterprise Fund (BRIEF) award, providing funding for research work to which this book is related.

Anthony Barrett

Dr Barrett has worked extensively in the field of Image Processing and has published around 30 papers including several books on the subject. Dr Barrett was an SERC grant holder for a joint project with industry in Image Processing from 1991–1994.

David Martland

David Martland lectures in the field of Neural Networks and Genetic Algorithms in the Department of Computer Science and Information Systems at Brunel University. He has developed a number of courses in these areas and presented corresponding papers at several International Conferences.

Contents

Contributors

Anthony Barrett
Department of Computer Science and Information Systems at St. John's
Brunel University
Uxbridge
Middlesex. UB8 3PH
United Kingdom

Deryn Graham
Department of Computer Science and Information Systems at St. John's
Brunel University
Uxbridge
Middlesex. UB8 3PH
United Kingdom

David Martland
Department of Computer Science and Information Systems at St. John's
Brunel University
Uxbridge
Middlesex. UB8 3PH
United Kingdom

Part 1
Knowledge-Based Systems: An Overview

D. GRAHAM and D. MARTLAND

1 Introduction to Knowledge-Based Systems

D. GRAHAM

No attempt is made here to embrace the subject of knowledge-based systems in its entirety given the limited space available. In this part of the book we aim only to provide the reader with an overview of knowledge-based systems, what they are, how they fit in within the broader area of artificial intelligence, some of the terminology, architectures and applications. Looking at some of the outstanding problems and issues, particularly with respect to knowledge elicitation and knowledge representation. A description is included of a significant, sub-discipline of artificial intelligence in relation to image processing and knowledge-based systems namely neural networks. Exercises and references to further reading are provided in Appendix A.

1.1 BACKGROUND TO KNOWLEDGE-BASED SYSTEMS

Knowledge-based systems and *Expert systems* are systems developed as a result of work in Artificial Intelligence. Artificial Intelligence (AI) is:

'. . . that part of computer science concerned with designing intelligent computer systems, that is, systems that exhibit the characteristics we associate with intelligence in human behaviour – understanding language, learning, reasoning, solving problems, and so on.'

(Barr and Feigenbaum, 1981)

Artificial Intelligence can be decomposed into a number of sub disciplines such as Game Playing, Machine Learning, Natural Language Processing, Vision, Robotics, Neural Networks and Parallel Distributed Processing (PDP), as well as Expert or Knowledge-Based Systems. (Neural Networks is of special significance to Knowledge-Based Systems and Image Processing, and will therefore be expanded upon in Chapter 5). This list is by no means exhaustive and decompositions vary among authors.

Some important features, common to all composites of the field of AI (Luger and Stubblefield, 1993) are:

- that computers are used to perform symbolic reasoning;
- problems addressed do not respond themselves to algorithmic solutions, underlying the reliance on *heuristic* search as an AI problem solving technique;
- problem solving using inexact, missing, or defined information requiring representational formalisms to be employed that enable the programmer to compensate for these problems;
- an aim to capture and manipulate the significant qualitative features of a situation rather than relying on numeric (quantitative) methods;
- issues of semantic meaning as well as syntactic form addressed;
- 'sufficient' answers, neither exact or optimal, resulting from the essential reliance on heuristic problem-solving methods. Such methods employed in situations where optimal or exact results are either too expensive or not possible;
- employment of large amounts of domain specific knowledge in problem solving, the basis of expert systems;
- application of *meta-level* knowledge to effect more sophisticated control problem-solving strategies.

Before proceeding further one should elaborate on some of the terms introduced above. A *heuristic* is a 'rule of thumb', based upon prior knowledge and experience. For example, if your personal computer fails to work, you would not immediately consider every possible fault for the symptoms manifested, at a component or model-based level. You would more likely check first that the power is on, that all the connections are sound etc before ever contemplating the internal workings, i.e. you'd check the 'obvious', most likely problems based upon your knowledge and experience.

Meta-level knowledge is knowledge about knowledge. It relates more to structure and strategy, for instance, knowledge about when knowledge is to be applied, when or how rules should be fired.

Efforts to understand the nature of intelligent thought go back much farther than the digital computer and the advent of AI; there have been many centuries of philosophical activity. The two most fundamental concerns of AI are *knowledge representation* and *search*.

Knowledge representation addresses the problem of capturing the full range of knowledge required for intelligent behaviour in a formal language i.e. one suitable for computer manipulation. Knowledge representation will be expanded upon further in chapter 4, looking mainly at the knowledge representation schemes for knowledge-based systems.

Search (Green, 1984) is a problem-solving technique that systematically explores a space of problem states, namely, successive and

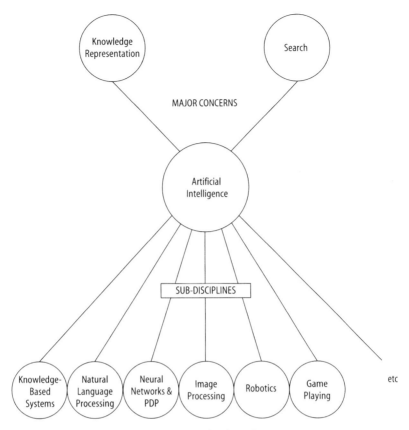

FIGURE 1.1.   Artificial Intelligence

alternative stages in the problem-solving process. Search strategies are usually embodied mostly within the inference mechanisms of knowledge-based systems and their respective knowledge representation schemes, for example, PROLOG. Hence we look at knowledge representation in a little more detail in this book, and therefore now only very briefly look at search.

In a problem solving program, we have:

(i) A representation scheme in which the problem, and the knowledge required for its solution, can be expressed;

(ii) A Knowledge-Base, constructed using the representation scheme – what we know about the problem domain;

(iii) Knowledge about the particular problem, encoded in a similar way;

(iv) An 'Inference Engine' or *Search method* which allows us to map (i) onto (ii).

In general, the more knowledge the less search e.g. chess, expert trouble shooting. Improved problem solving behaviour can usually be

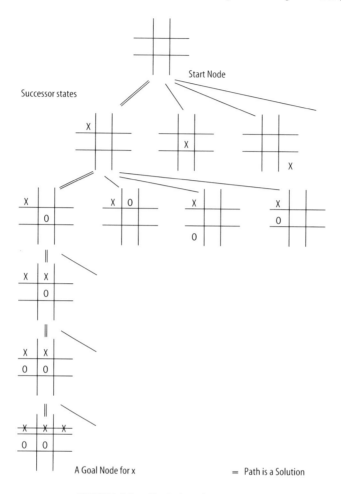

FIGURE 1.2. Exploring the State Space.

traced to extra knowledge, or a better scheme for encoding knowledge, rather than to improved search.

We need a way of organising the search, all knowledge-based systems require this. Early AI placed great emphasis on search, modern AI emphasises representation and knowledge.

1.1.1 Search Strategies

Search strategies are employed to solve problems by exploring *likely consequences of potential actions*. The exploration takes place in state-space, the space of allowed problem states. *Arcs* in state-space connect those *nodes* (individual problem states) which can be reached by a single legal action. Figure 1.2 gives a very simple example for the game of Noughts and Crosses, with some (not all) of the successor states shown.

A search-based problem solver makes decisions based on the results of the *exploration of the likely consequences of potential actions*. Rather than explicitly representing the knowledge needed for solving problems in a given domain, mechanisms are provided for conducting a search. The following is some of the terminology used:

- *State-space* is the space of allowed problem states;
- *Nodes* in state-space represent individual problem states;
- *Arcs* connect a node to those other nodes which can be reached by a single legal move;
- *An Operator* is a procedure embodying a means of transforming one state into another – a move;
- *The Legal Move Generator* is a procedure which, given a node, returns all those nodes which can be reached by the application of an operator. It embodies the 'rules' of the problem;
- *The Start Node* is the initial problem state;
- *A Goal Node* is a problem state representing a solution. There may be more than one goal node;
- *A Solution* is a path from the Start Node to a Goal Node.

It may be sufficient to find *any* solution. Alternatively, the '*best*' solution may be required. In a simple case, this might be the *shortest path*. More generally, we can have '*costs*' associated with each arc. Now the best solution is that with minimum total cost. (Map Traversal is an example of non uniform costs).

State-space may take the form of a *tree*, or, when it is possible to return to a previously visited state, a *graph*. In all but trivial cases it is *not possible to explore state-space fully* (i.e. until every path reaches a goal state, or a dead end). If the *branching factor*, the number of successors to a given state is b and the tree is explored to a depth N, there will be b^N nodes at the Nth level, this is known as the *combinatorial explosion*.

To program a state-space search, we need:

- A way of representing the states;
- A way of representing the operators and hence of programming the legal move generator;
- A way of specifying the start of goal states, and of recognising a goal state;
- A search strategy.

It should be noted that if state-space is a graph, search may never terminate.

Ideally search strategies should be:

(i) Guaranteed to find a solution if one exists;
(ii) Guaranteed to find the '*best*' solution (i.e. that which represents the least cost or, if costs are uniform, the solution of minimum depth from Start). Such an algorithm is said to be *admissible*;

(iii) Guaranteed to find the best solution *efficiently* i.e. with little or no wasted effort. The most efficient admissible algorithm is said to be *optimal*.

Efficiency:

Number of nodes on solution path / Number of nodes on closed < 1 (1 = OPTIMAL solution).

Several strategies for conducting search exist, such as:

(i) Breadth-First Search;
(ii) Depth-First Search;
(iii) Modified Depth-First Search to a Depth Limit;
(iv) Knowledge-Guided Depth-First Search;
(v) Best-First Search (A^T Search);
(vi) A* Search;
(vii) Beam Search.

Search Strategies will not be elaborated upon further here.

1.2 WHAT IS A KNOWLEDGE-BASED SYSTEM?

Firstly, let us distinguish between knowledge-based systems and expert systems (Graham, 1990).

Knowledge-based systems enhance, through providing support, that which a human agent (often expert) does. The computer support is likely to consist of some representation of the agent's knowledge, typically rules, with an inference mechanism, plus or minus some uncertainty handling features (Johnson, 1990).

An *expert system* is a knowledge-based system with an evaluated level of performance close to that of an expert. Knowledge-based systems and expert systems are often not distinguished i.e. if an expert is involved, then usually the implication is that it is an expert system (Johnson, 1990). The term knowledge-based system is sometimes used as a synonym for 'expert system'. Knowledge-based systems may be considered to be a generic term which encompasses expert systems. Feigenbaum (Harmon and King, 1985, p. 5) defines an expert system as:

'. . . an intelligent computer program that uses knowledge and inference procedures to solve problems that are difficult enough to require significant human expertise for such a level, plus the inference procedures used, can be thought of as a model of expertise of the best practitioners of the field.'

A more recent definition (Jackson, 1990, p. 3):

> 'An expert system is a computer program that represents some specialist subject with a view to solving problems or giving advice.'

Feigenbaum (Harmon and King, 1885, p. 5) also states that:

> 'The knowledge of an expert system consists of facts and heuristics. The "facts" constitute the body of information that is widely shared, publicly available, "heuristics" are mostly private, little-discussed rules of good judgement (rules of plausible reasoning, rules of good guessing) that characterise expert-level decision making in the field. The performance level of an expert system is primarily a function of the size and quality of a knowledge base it possesses.'

The terms 'knowledge engineers' and 'knowledge engineering' are also attributable to Feigenbaum, to describe the people who build knowledge-based expert system and the technology. Knowledge engineering is considered to be 'applied artificial intelligence' (Feigenbaum, 1977).

1.2.1 Basic Architecture

The diagram below depicts the basic architecture of a knowledge-based or expert system. The two main components are the knowledge base and the inference engine. The former embodies the expert's knowledge of a particular domain, the latter contains inference strategies used by an expert when manipulating the knowledge base. The knowledge base is made-up of facts and rules. Various representation schemes exist for encoding these facts and rules, such as production rules (Davis and King, 1977), structured objects (Findler, 1979) and predicate logic (Kowalski, 1979). Most expert systems use one or more of these formalisms, commonly employing a combination of rules and frames.

The *inference engine* is the module that uses the knowledge base coupled with information elicited from the user during a particular session in order to draw conclusions. Thus the knowledge base contains information *specific to the problem domain*, whereas the inference engine is *generic to any problem*.

The inference mechanism may be backward- or forward-chaining. Backward-chaining; the program reasons backward from what it wants to prove towards the facts that it needs. Forward-chaining; reasoning forward from the facts that it possesses. Searching for a solution by backward-chaining is generally more focused than searching by forward-chaining, because one only considers potentially relevant facts.

The *working memory* contains conclusions specific to the current session, derived from information elicited from the user, this information is known as *inferred knowledge*, and is not part of the overall knowledge base.

The *user interface* is a part of the system that is most variable. In traditional systems it took the form of a menu driven text based interface. In the drive to make expert systems more effective and user friendly today's user interfaces may be via speech recognition, graphical user interfaces (GUI) or even via another third party application. In which case the expert system is said to be *embedded*, and has no direct interaction with the user.

Most expert systems also include some sort of *rule editor* that is able to update the knowledge base. This may be done on an 'as required' basis by a knowledge engineer, or may be performed automatically and drive by interfaces to other applications such as relational databases.

MYCIN (Shortliffe, 1976) was the first expert system that was able to provide an explanation for its conclusions. This feature is provided by a *solution rationale module* (sometimes known as an *explanation subsystem*).

1.2.2 Characteristics of an Expert System

An expert system (Harmon and King, 1985; Jackson, 1990, pp. 4–5) is distinguishable from a more conventional applications program in that:

- It simulates human reasoning about a problem domain;
- It performs reasoning over representations of human knowledge, in addition to doing numerical calculations or data retrieval;
- It solves problems by heuristic or approximate methods which, unlike algorithmic solutions, are not guaranteed to succeed.

Expert systems encode the domain-dependent knowledge of everyday practitioners in some field, and use this knowledge to solve problems, instead of using comparatively domain-independent methods derived from computer science or mathematics. An expert system differs from other kinds of artificial intelligence program in that:

- It deals with subject matter of realistic complexity that normally requires a considerable amount of human expertise;
- It must exhibit high performance in terms of speed and reliability in order to be a useful tool;
- It must be capable of explaining and justifying solutions or recommendations to convince the user that its reasoning is in fact correct.

1.2.3 Explaining Solutions

Explanations of expert system behaviour are important for a number of reasons (Jackson, 1990, pp. 11–12):

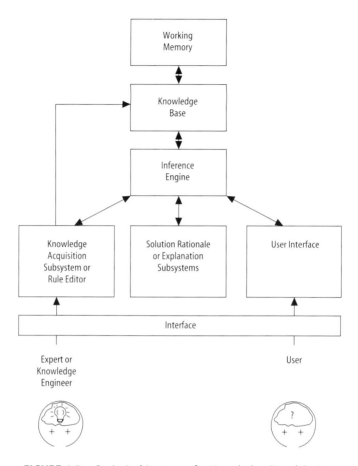

FIGURE 1.3. Basic Architecture of a Knowledge-Based System

- Users of the system need to satisfy themselves that the program's conclusions are basically correct for their particular case;
- Knowledge engineers need some way to satisfy themselves that knowledge is being applied properly even as the prototype is being built;
- Domain experts need to see a trace of the way in which their knowledge is being applied in order to judge whether knowledge elicitation is proceeding successfully;
- Programmers who maintain, debug and extend knowledge-based programs must have some window on the program's behaviour above the level of the procedure call;
- Managers of expert system technology, who may end up being responsible for a program's decisions, need to satisfy themselves that a system's mode of reasoning is applicable to their domain.

Explanation sometimes goes under the name of transparency, the ease with which one can understand what the program is doing and why. Closely linked to the topic of evaluation, explanation allows the scrutinisation of the outputs of a system and examination of a trace of its reasoning, thus enabling decisions as to whether or not that system is getting the right answer for the right reasons.

1.2.4 Uncertainty

Knowledge-based or expert systems are designed to interface with the real, non-deterministic world. They do so by allowing for uncertainty. There are two forms of uncertainty which effect expert systems (Palmer, 1995):

- Uncertainty as to the validity of a particular rule;
- Uncertainty as to the validity of data entered by the user.

Thus expert systems must be able to take into account all the associated certainty factors when reaching a conclusion. Uncertainty within an expert system is usually expressed in terms of a confidence factor (cf). In general, $0 \leq cf \leq 1$, where 0 is false and 1 is completely true (other systems do exist, using different ranges and negative values). The confidence factor associated with a particular rule is determined during the knowledge elicitation stage of the process. It is by no means a scientific process and is usually based on the judgement of the domain expert. The confidence factor associated with a particular piece of knowledge is usually provided by the user at run time. In order for the expert system to deal with these factors it must be able to provide a resultant confidence factor for the following situations involving fact a and fact b:

> a and b
> a or b
> not a
> if a then <conclusion>

The way in which expert systems deal with these situations varies.

Uncertainty may be considered as inadequate information to make a decision. It can occur as a result of errors, due to ambiguity, incompleteness, or incorrectness. Theories that have been developed to overcome this include classical and Bayesian probability (as used in MYCIN), Zadeh's fuzzy theory and Dempster-Shafer theory (Garcia, 1991). In Bayesian probability, the probabilities of previous known results are used to calculate more complex probabilities. This latter approach has a number of limitations; it requires updating and is limited in use by humans for problem solving (HMSO, 1991). It can be seen that although these rules are intuitive, they obviously are not deduced by quantitative means. This is now changing; Krause and Clark (1993) examine recent research into more quantitative methods of determining uncertainty. These methods include fuzzy sets, possibility theory, non-monatomic logic and augmentation. Some systems require inexact reasoning with uncertain facts or rules. MYCIN and PROSPECTOR are well documented examples of early systems which deal with uncertainty.

1.2.5 Evaluating Expert Systems Generally

Jackson (1990, p. 55) suggests a number of preconditions are necessary for evaluation to be meaningful:

- There must be some objective criteria for success;
- Proper experimental procedures must be followed;
- Evaluation should be done painstakingly or not at all;
- Finally, some expert systems may play very different roles in problem solving, which require rather different standards of performance.

2 Knowledge-Based System Architectures and Applications

D. GRAHAM

2.1 KNOWLEDGE-BASED SYSTEM TOOLS AND ARCHITECTURES

Five main areas where expert systems are being successfully applied in the commercial market place (Aryanpur, 1988), are:

- diagnostic problems;
- materials selection;
- advisory systems;
- intelligent front ends for complex programs;
- risk analysis.

Expert system development tools on the market vary in sophistication and the complexity of expert systems they are able to produce.

Present day expert systems are unlikely to be constructed from scratch. Usually, some form of tool will be employed. The term tool encompasses: programming environments, which provide intelligent editors and various knowledge engineering constructs, as well as the constructs provided by an ordinary symbolic manipulation language; knowledge representation languages embedded on one or more knowledge representation schemes, notable examples include KRL (Bobrow and Winograd, 1977) and KRYPTON (Brachman *et al.*, 1983); and shells. The majority of software tools for building expert systems seem to fall into five broad categories (Jackson, 1990, pp. 338–339):

(1) *Expert system shells*, which are essentially abstractions over one or more applications programs;
(2) *High-level* programming languages, which to some extent conceal their implementation details, thereby freeing the programmer from low-level considerations of efficiency in the storage, access and manipulation of data. OPS5 (Forgy, 1982) is a good example of such a language;

(3) *Multiple-paradigm programming environments*, sometimes called *hybrid systems*, which provide a set of software modules that allow the user to mix a number of different styles of artificial intelligence programming. Example: LOOPS (Bobrow and Stefik, 1983);

(4) *Skeletal systems*, which provide the knowledge engineer with a basic problem-solving program to be instantiated. Examples: BB* (Hayes-Roth *et al.*, 1987) and ABE (Erman *et al.*, 1986). BB* is a '*blackboard architecture*', and ABE is a design for a software tool which facilitates programming with large communicating modules of code embedded in an operating system. Such programs are meant to provide the user with an environment in which it is relatively easy to select, elaborate and combine skeletal systems;

(5) *Additional modules* such as dependency networks for performing specific tasks within a problem solving architecture. An example of this kind of module is the dependency network used by the expert system VT, to keep track of which values of design variables depend on values determined by earlier decisions. These networks can also be used to propagate updates brought about by changing data or assumptions, in which case they are called *Truth Maintenance Systems* (TMSs).

2.1.1 Expert System Shells

A shell or skeletal system, is defined by Keravnou and Johnson (1986) as being:

'*. . . a generalisation of an expert system, made by deleting the domain specific knowledge from the knowledge base and adding the facilities necessary for instantiating the knowledge base for some other domain.*'

Early expert systems were built 'from scratch', in the sense that the architects either used the primitive data and control structures of an existing programming language to represent knowledge and control its application, or implemented a special-purpose rule or frame language in an existing programming language, as a prelude to representing knowledge in that special-purpose language. The special-purpose languages typically had two different kinds of facility:

- modules, such as rules or frames, for representing knowledge; and
- an interpreter, which controlled when such modules became active.

The modules, taken together, constituted the knowledge base of the expert system, while the interpreter constituted the inference engine. In some cases, it was clear that these components were reusable, in the sense that they would serve as a basis for other applications of expert systems technology. Because such programs were often abstractions of existing expert systems, they became known as expert system shells.

A classic example of this is EMYCIN (vanMelle *et al.*, 1981; vanMelle, 1979) which was a domain-independent framework for constructing and running consultation programs. EMYCIN ('Empty' or 'Essential' MYCIN) was derived from MYCIN (see section 2.2.1) and provided a number of features which have since become widespread in expert system shells (Jackson, 1990, pp. 224–225):

- An abbreviated rule language;
- An indexing scheme for rules. Based on the parameters they reference, this also organises the rules into groups;
- A backward-chaining control structure like MYCIN's which unfolds an AND/OR tree, the leaves of which are data that can be looked up in tables or requested from the user;
- Interfaces between the final consultation program and the end-user, and the system designer and the evolving consultation program. The interface between the final consultation program and the end-user handles all communications between the program and the user (for example, the program's requests for data and provision of solutions; the user's provision of data and requests for explanations);
- Interface between the system designer and the evolving consultation program. This provides tools for displaying, editing and partitioning rules, editing knowledge held in tables, and running rule sets on sets of problems.

The main advantages of shells, their simplicity and generality, are also their main disadvantages. Many criticisms have been voiced concerning the inflexibility of shells (Aikins, 1983; Alvey, 1983; Meyers *et al.*, 1983). To redress the balance a little, shells can be useful for prototyping and gaining basic knowledge of knowledge-based systems i.e. educational benefits.

2.1.2 High-Level Programming Languages

Production rule languages, object-oriented programming languages, and procedural deduction systems fall into this category. They provide programmers with a fast prototyping tool and more freedom than a shell in relation to control (e.g. forward- or backward-chaining) and uncertainty handling.

Although a shell may be considered more user friendly, the advantage is flexibility. This is of particular importance for experimental system development where the mechanisms required to solve the domain may not be apparent (Jackson, 1990, pp. 342–3).

2.1.3 Hybrid Systems

Known also as multiple-paradigm programming environments, hybrid systems combine different software modules and representational schemes, in an attempt to overcome the individual weaknesses of single schemes of representation and inference.

Early attempts, married rule- and frame-based styles in an endeavour to replace the procedural obscurity of demonic attachments with the clarity of productions. Production rules (see section 4.4) also being set in a context provided a richer representation of the data than that afforded by working memory.

The first AI programming environment to mix four programming paradigms (within a message-passing architecture) was LOOPS (Bobrow and Stefix, 1983). It combined procedure-oriented programming with rule-oriented programming, object-oriented programming and data-oriented programming.

Later environments such as KEE (Intellicorp, 1984; Kunz *et al.*, 1984) and Knowledge Craft (Carnegie Group, 1985) surmounted many of the problems raised by LOOPS. KEE possesses features such as multiple inheritance of properties, message-passing (replacing procedure writing) and the rules language more closely resembling the syntax and semantics of productions, are supported. The package is embedded in a congenial programming environment that uses high-resolution graphics to display object hierarchies and windows representing individual objects.

Both KEE and Knowledge Craft also have an added query language in the PROLOG style to the paradigms listed above. The object-oriented paradigm is central to such systems, with different programming styles attaining varying degrees of integration within this. KEE allows the behaviour of an object to be described in terms of a set of production rules, Knowledge Craft allows procedures attached to objects to initiate, or return control to both the production rule and the logic programming modules. KEE and Knowledge Craft also have implementations for VAX in addition to running on LISP machines.

Such systems allow knowledge engineers to create their own architecture, using the building blocks provided and the interfaces between them. This facilitates a great deal of flexibility but the task is non-trivial, and few tools currently provide any on-line assistance (Jackson, 1990, pp. 348–350).

2.1.4 Skeletal Systems

Blackboard systems (Jackson, 1990, pp. 368–369) outlined below, can emulate both forward- and backward-chaining reasoning systems, as well as being able to combine these forms of reasoning opportunistically. They provide the basis for an abstract architecture of great power and generality. The blackboard model of problem solving encourages the hierarchical organisation of both domain knowledge and the space of partial and complete solutions. It is therefore well suited to construction problems in which the problem space is large but factorable along some number of dimensions. Successful applications of the blackboard approach have included data interpretation (such as image understanding, speech recognition), analysis and synthesis of complex compounds (such as protein structures) and planning.

A blackboard system is organised essentially with the domain knowledge partitioned into independent *knowledge sources* (KSs) which run under a *scheduler*. Solutions are thus built up on a global data structure, the *blackboard*. Instead of representing 'how to' knowledge in a single rule set, such as is encoded in a suite of programs, each of these knowledge sources may be a rule set in its own right, or the suite may contain diverse programs that mix rules with procedures.

Rather like that of a production system's working memory, except that its structure is much more complex, the blackboard now serves this function. The blackboard is typically partitioned into different *levels* of description with respect to the solution space, each corresponding to a different amount of detail or analysis. Levels usually contain more complex data structures than working memory vectors, such as object hierarchies or recursive graphs. In modern systems, there may be more than one blackboard.

Although knowledge sources trigger on blackboard objects, their instantiations may not be executed at once. *Knowledge source activation records* (KSARs) are normally placed on an agenda pending selection by the scheduler instead. Knowledge sources communicate only via the blackboard, and thus cannot transmit data to each other or invoke each other directly. This is similar to the convention in production systems that forbids individual rules from calling each other directly; everything must be done through working memory.

2.1.5 Truth Maintenance Systems

Mechanisms for keeping track of dependencies and detecting consistency are often referred to as *truth maintenance systems* (TMSs), occasionally called *reason maintenance systems*. The function of a truth maintenance component in the context of a larger problem solving program is to (Jackson 1990 p. 397):

- cache inferences made by the problem solver, so that conclusions that have once been derived need never be derived again;
- allow the problem solver to make useful assumptions and see if useful conclusions can be derived from them; and
- handle the problem of inconsistency, either by maintaining a single consistent world model, or by managing multiple contexts which are internally consistent but which may be mutually inconsistent.

TMSs (e.g. such as that of Doyle and McAllester) are useful for finding a single solution to a constraint satisfaction problem. The use of a dependency net in the VT expert system, is an excellent example of how a simple TMS can be used to good effect. A single solution to the problem is required; we are not really interested in generating the whole range of alternative designs. If more than one solution, or all solutions are required, extra control machinery is needed. Because dealing all the time with a single state of the network, it makes it difficult to compare alternative solutions to the problem, if this is really necessary. In differential diagnosis, for example, it is usually important to compare hypotheses with their competitors to arrive at a composite hypothesis which covers the data and gives the best account of it.

Assumption-based TMS (ATMS) (De Kleer, 1986) is more oriented towards finding all solutions to constraint satisfaction problems, finding all the interesting environments that satisfy the constraints. Extra control machinery is required to find fewer solutions, or a single 'best' solution according to some criteria. However, since more than one context can be maintained, this obviates the need for dependency-directed backtracking in the basic version.

2.1.6 Current Research Problems in Expert Systems

Current research problems that impinge on practical aspects of expert systems include: *diagnosis from first principles, inexact reasoning* and *rule induction*. These problems have implications for the central problem of knowledge acquisition (Jackson, 1990, pp. 466–467).

Diagnosis from first principles suggests that it may be possible to side-step the whole business of deriving fault-finding heuristics for some device from a diagnostic expert. Deriving diagnosis of faulty behaviour by reasoning about minimal perturbations to a representation of the device's *correct* behaviour. Because it is more firmly based in an explicit model of the domain than most rule-based approaches, this approach to diagnosis is sometimes called 'model-based'. Jackson (1990, p. 467) states that there are however, problems with this approach, such as:

- *Getting a system description in the first place.* In some domains, such as troubleshooting electronic circuits, it is possible that the description could be derived as a by-product of computer-aided design. Certainly this description must be complete and consistent for the method to work properly;
- *Controlling the combined explosion of multiple fault hypotheses.* This problem can be rendered tractable by information-theoretic methods if we make various independence assumptions about component failures. However, in a domain where these assumptions do not hold, or where we also need to consider the failure of conditions between components, the combinatorial problems reappear.

Inexact reasoning research is proposing ways in which belief update can be more closely tied to the structure of the problem. Dempster-Shafer theory and Bayesian belief updating can thus both be adopted for modelling probabilistic inference in hierarchical hypothesis spaces. Another way of tying inexact methods to deeper knowledge of the domain is suggested by the link between Bayesian methods and causal networks, instead of attaching certainties of more superficial associations between causes and effects. With a model of hierarchical or causal relationships among data and hypotheses it may make the assignment of weights of evidence to hypotheses easier than the assignment of certainties to the conclusions of individual decision rules. The explicit model to hand may encourage the expert and the knowledge engineer to devise and enforce a coherent policy on the assignment of weights, and help the knowledge engineer to design a knowledge-acquisition program to check that this policy is consistently applied, for example.

Research on rule induction is providing a methodology for deriving, evaluating and tuning sets of decision rules. Such methods appear to be required if really large knowledge bases are to be built and maintained. Machine learning also, is fundamental to the whole area of 'self-improving' systems, which gain in knowledge from their problem-solving experiences. Techniques which are of direct relevance to expert systems applications are:

- Learning from examples; programs have been built which are capable of learning concepts from examples; constructing rules which correctly classify a set of positive and negative training instances, and then using these rules to classify unseen data with a high degree of accuracy;
- Application of decision theory to the evaluation of individual heuristic rules and the optimisation of sets of such rules, with a view to improving the quality of rule sets in terms of both analysability and performance.

Inductive learning research has made great progress, and is certain to contribute to future knowledge engineering methodologies. Research on the evaluation of heuristic rules seeks to demystify the art of encoding expertise in rules by showing to what extent rule weights can be varied before the output of a program is affected. Knowledge elicitation will be facilitated by greater understanding in this area, and lead to more robust systems.

2.1.7 Professional Issues

Jackson (1990) among others, points out that there are many issues to do with the rights of people affected by this technology that need to be addressed. He suggests that professional bodies should have an educative role to play in this respect, promoting public awareness and encouraging informed debate in an atmosphere somewhat removed from political pressures and commercial considerations. They also have a role to play in maintaining standards in AI-related products and developing codes of practice for the deployment of AI technology.

2.2 KNOWLEDGE-BASED SYSTEM APPLICATIONS

There are now numerous knowledge-based system or expert system applications in existence, too many to elaborate upon. Instead, we will focus our discussion on medical systems.

2.2.1 Early Systems

One of the earliest and perhaps most significant systems, was MYCIN (Jackson, 1990), for which work began in 1972. Shortliffe (1976) gives the most complete account of this work. MYCIN's purpose is to assist a physician who is not an expert in the field of antibiotics with the treatment of blood infections. Work on MYCIN began as a collaboration between the medical and AI communities at Stanford.

Any drug designed to kill bacteria or arrest their growth is known as an 'anti microbial agent'. There is no single agent effective against all bacteria and some agents are too toxic for therapeutic purposes. The selection of therapy for bacterial infection is in four-parts, determining:

(1) if the patient has a significant infection;
(2) the (possible) organism(s) involved;
(3) a set of drugs that might be appropriate;
(4) the most appropriate drug or drug combination.

Samples taken from the site of infection are sent to a microbiology laboratory for culture. Evidence of organisms grown from these samples may allow a report of the morphological or staining characteristics of the organism. The range of drugs any identified organism may be sensitive to may be unknown or uncertain, however.

The basic version of MYCIN has five components:

(1) a knowledge base (containing domain knowledge);
(2) a dynamic patient database (containing individual case information);
(3) a consultation program;
(4) an explanation program;
(5) a knowledge acquisition program (for adding new rules and changing existing ones).

Components 1–3 is the system which comprises the problem-solving part of MYCIN, which generates hypotheses with respect to the offending organisms, and makes therapy recommendations based on these hypotheses.

MYCIN's knowledge base is organised around a set of rules of the general form:

- if condition 1 and . . . and condition m hold
- then draw conclusion 1 and . . . and conclusion n

encoded as data structures of the LISP programming language. These if . . . then statements are not program statements of the LISP language, they are declarations of knowledge which will be interpreted by the consultation program (which is a LISP program). For example:

IF: (1) The stain of the organism is Gram negative; and
    (2) The morphology of the organism is rod; and
    (3) The aerobicity of the organism is aerobic;

THEN: There is strongly suggestive evidence (0.8) that the class of the organism is Enterobacteriacea

The conditions of a rule can also be satisfied with varying degrees of certainty, e.g. :

> if condition 1 holds with certainty x1 . . . and condition m holds with certainty xm then draw conclusion 1 with certainty y1 and . . . and conclusion n with certainty yn

where the certainty associated with each conclusion is a function of the combined certainties of the conditions and the tally, which is meant to reflect our degree of confidence in the application of the rule. The (0.8) being the cf for the conclusion of enterobacteriacea above, for instance.

A rule is a premise-action pair; such rules are sometimes called 'productions' for purely historical reasons. Premises are conjunctions of conditions, and their certainty is a function of the certainty of these conditions. Conditions are either propositions which evaluate to truth or falsehood with some degree of certainty, for example:

> the organism is rod-shaped

or disjunctions of such conditions. Actions are either conclusions to be drawn with some appropriate degree of certainty (such as the identity of some organism) or instructions to be carried out (such as compiling a list of therapies).

In addition to rules, the knowledge base also stores facts and definitions in various forms:

- simple lists, such as the list of all organisms known to the system;
- knowledge tables, which contain records of certain clinical

parameters and the values they take under various circumstances, such as the morphology (structural shape) of every bacterium known to the system;

- a classification system for clinical parameters according to the context in which they apply, such as whether they are attributes of patients or organisms.

MYCIN's control structure uses a top-level goal rule which defines the whole task of the consultation system, which is paraphrased below:

IF   (1) there is an organism which requires therapy; and
      (2) consideration has been given to any other organisms requiring therapy;

THEN compile a list of possible therapies, and determine the best one in this list.

A consultation session follows a simple two-step procedure of creating the patient context as the top node in the context tree, and attempting to apply the goal rule to this patient context.

MYCIN can ask for items of data as and when they are needed for the evaluation of a condition in the current rule, but does not enter an extended dialogue with the user. If the required information is not forthcoming, there may be rules which apply to the clinical parameter in question and which can be invoked in search of its value, but if there are no such rules the current rule application will fail.

This control structure is quite simple as AI programs go:

(1) The subgoal set up is always a generalised form of the original goal;
(2) Every rule relevant to the goal is used, unless one of them succeeds with certainty.

Evaluation of MYCIN

There are many ways in which to evaluate an expert system, the most obvious one is in comparison with a human expert. In developing the system, expert and knowledge engineer typically work together on a set of critical examples until the program can solve them all. Evaluation then involves giving the system 'unseen' examples and seeing if its judgements agree with those of the expert. In MYCIN's case, the program is attempting to mimic some of the judgmental reasoning of a human expert, and so comparison with the expert is relevant.

MYCIN was never used in hospital wards for a number of reasons, including:

- Its knowledge base is incomplete, since it does not cover anything like the full spectrum of infectious diseases;

- Running it would have required more computing power than most hospitals could afford at that time;
- Doctors do not relish typing at the terminal and require a much better user interface than that provided by MYCIN in 1976.

MYCIN was a research vehicle, and therefore did not set out to achieve practical application or commercial success. Nevertheless, descendants of MYCIN have seen application in hospitals. Also, the success of the 'subgoaling' approach has not been confined to medical applications.

2.2.2 Current Systems

As previously mentioned an expert system shell, EMYCIN, or Essential Mycin, was derived from MYCIN. PUFF was a system built using EMYCIN as a base. PUFF can interact with laboratory instruments for measuring tests on respirations given to patients in a pulmonary function laboratory (Edmunds, 1988). The tests determine the presence and degree of lung disease. PUFF also uses backward chaining strategies. INTERNEST (Pople *et al.*, 1977), encompasses the entire internal medicine field. Many difficult cases have been resolved using this. It was introduced in 1974, written in InterLisp. Its limitations have resulted in the development of CADUCEUS.

More recent applications and prototypes in the medical profession include GENISIS in 1982 (Fox and Rawlings, 1992), an integrated knowledge-based system for genetics to use for laboratory experiment plans and data management, using a frame structured knowledge and specialised languages. The authors define another system, PAPAIN (1990) which allows for automation of protein sequence and structure prediction with deductive databases and constraint logic programming techniques. A more recent application is LEPDIAG (Banerjee *et al.*, 1994), using a knowledge-based system for diagnosis and monitoring of leprosy. A fuzzy tool called FRUIT is used to overcome uncertainty, with fuzzy production rules, and metarules to resolve rule conflict. Other medical applications are AI/COAG for diagnosing blood disease, and ONCOCIN to remedy and manage chemotherapy patients. DentLE (Eisner *et al.*, 1993) and IMR-Entrt (Naeym-Rad *et al.*, 1993) are also examples of medical systems, the former uses hypermedia technology in a dentistry system, and the latter can be used in hospitals.

RaPiD (Hammond *et al.*, 1993) is a system developed as a design assistant for use in prosthetic dentistry. RaPiD integrates computer-aided design, knowledge-based systems and databases, employing a logic-based representation as the unifying medium. The user's manipulation of icons representing the developing design is interpreted as a set of transactions on a logic database of design components. The rules of design expertise are represented as constraints. When design rules are contravened as the result of some proposed alteration, a suitable critique is presented to the user. RaPiD is being developed for use in both dental education and practice.

3 Knowledge Elicitation

D. GRAHAM

3.1 KNOWLEDGE ACQUISITION AND KNOWLEDGE ELICITATION

Knowledge acquisition has been described as:

'the transfer and transformation of potential problem-solving expertise from some knowledge source to a program'

(Buchanan *et al.*, 1983)

The term knowledge acquisition is generic, it is neutral with respect to how the transfer of knowledge is achieved. Knowledge elicitation however, often implies that the transfer is accomplished by a series of interviews between a domain expert and a knowledge engineer who then writes a computer program representing the knowledge (or gets someone else to write it). It involves (Jackson, 1990, pp. 219–220): the elicitation of knowledge from experts in some systematic way – for example, by presenting them with sample problems and eliciting solutions; storing the knowledge so obtained in some intermediate representation; and compiling the knowledge from the intermediate representation into a runnable form, such as production rules.

Only two to five production rule equivalents per day are generated by a knowledge elicitation interview. It is suggested by Jackson (1990, p. 220) that productivity is so poor due to:

- 'The technical nature of the specialist fields requires the non-specialist knowledge engineer to learn something about the domain before communication can be productive;
- Experts tend to think less in terms of general principles and more in terms of typical objects and commonly occurring events;
- The search for a good notation for expressing domain knowledge (some form of mediating representation scheme), and a good framework for fitting it all together, is itself a hard problem, even before one gets down to the business of representing the knowledge in a computer.'

The process of knowledge acquisition is often broken down into sub tasks that are easier to understand and simpler to carry out.

3.2 STAGES OF KNOWLEDGE ACQUISITION

Several stages of knowledge acquisition are suggested in the literature e.g. Buchanan *et al.* (1983), who offer an analysis of knowledge acquisition in terms of a process model of how to construct an expert system. Wielinga and Breuker (1985) distinguish between five different levels of analysis:

- knowledge identification;
- knowledge conceptualisation;
- epistemological analysis;
- implementation analysis;
- ontological analysis.

Ontological analysis (Alexander *et al.*, 1986), describes systems in terms of entities, relations between them, and transformations between entities that occur during the performance of some task. The authors use three main categories for structuring domain knowledge:

- static ontology;
- dynamic ontology;
- epistemic ontology.

There is some obvious overlap with the knowledge conceptualisation and epistemological analysis levels of Wielinga and Breuker's framework. However, there is less of a correspondence with lower levels, such as the logical and implementational analyses. Ontological analysis assumes that the problem under study can be reduced to a search problem, but does not focus on the method of search. These analyses may seem rather abstract, but they are valuable because they help to structure an ill-structured task. Essentially all analyses embody common notions as depicted by Figure 3.1 below.

Identify and Conceptualise: Identify the class of problems that the system will be expected to solve, including the data that the system will work with, the criteria that solutions must meet and the resources available for the project, in terms of expertise, manpower, time constraints, computing facilities and money. Key concepts and the relationships between them; a characterisation of the different kinds of data, the flow of information and the underlying structure of the domain, in terms of causal, spatiotemporal, or part-whole relationships, and so on.

Formalise: Attempt to understand the nature of the underlying search space, and the character of the search that will have to be

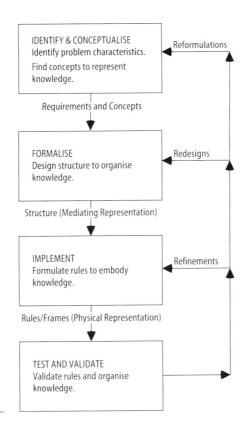

FIGURE 3.1
Fundamental Stages of
Knowledge Acquisition

conducted. Important issues include the certainty and completeness of the information, and other constraints on the logical interpretation of the data, such as time dependency, and the reliability and consistency of different data sources. A mediating representation (between the elicited form e.g. interviews, and the implemented form e.g. production rules) results from this stage, a 'paper model'.

Implement: In turning a formalisation of knowledge into a runnable program, the primary concern is with the specification of control and the details of information flow. Rules will have to be expressed in some executable form under a chosen control regime, while decisions must be made about the data structures and the degree of independence between different modules of the program.

Test and Validate: The evaluation of expert systems is far from being an exact science, but it is clear that the task can be made easier if one is able to run the program on a large and representative sample of test cases. Common sources of error are rules which are either missing, incomplete or wholly incorrect, while competition between related rules can cause unexpected bugs.

The primary consideration in designing an expert system is the class of problems that you want the system to solve.

3.3 KNOWLEDGE ACQUISITION METHODOLOGIES

Knowledge acquisition involves the elicitation of the data from the expert (usually via some verbal technique); the interpretation of the data to infer the underlying knowledge reasoning process, and, creation of a model of the expert's domain knowledge and performance (Kidd, 1987). These three phases have been similarly identified elsewhere in the literature as domain definition, fundamental knowledge formulation, and basal knowledge consolidation (Grover, 1983), or orientation, problem identification and problem analysis (Breuker and Wielinga, 1987a). They approximate also to: broad and shallow survey; task analysis, and; verification phases (Johnson and Johnson, 1987a). The common thrust is towards the development of a methodology of knowledge acquisition (Neale, 1988).

Until recently, comparatively little attention has been paid to documenting knowledge acquisition methodology, and authors have offered even fewer clues to the modelling of knowledge. There seems in many cases, implicitly or explicitly, to have been a move directly from the data to implementation by encoding the knowledge in IF . . . THEN rules.

The crucial intermediate stage of formulation of a conceptual model of the expert's knowledge has thus been omitted. This is important not least for reasons of expert validation: there must – if the system is intended to be a genuine attempt at emulation of the expert's problem solving expertise – be agreement between expert and knowledge engineer that the latter has properly and adequately acquired a knowledge of the concepts involved and their interrelationships, as used by the expert. In approaching knowledge elicitation through any of the various techniques (techniques will be discussed in the next section), the knowledge engineer should give a high priority to the documentation of the data and interpretation. The process should be foreseen as essentially incremental, operating in a breadth-first and cyclical manner to favour successive refinement and inclusion of the results of analysis of related topics (Breuker and Wielinga, 1987a). The acquisition of new data should alternate with analysis of data previously obtained. Documentation in the form of a 'paper model' in plain English is a useful form of feedback for expert review. Rather than representing a collection of production rules, which is generally an alien representation to most experts, the achievement of the proper level of understanding is more likely to be gained by examining together a map of the knowledge on paper first. This paper model might incorporate network or hierarchical forms of entity relationships, including inheritance features or weighted links; other graphical forms such as decision trees, and assertions and conditional statements. Diagrammatical or graphical representations, such as decision trees, are often more comprehensible to the expert than textual statements or a body of rules (see Alvey and Greaves (1987) for an example of the use of decision trees in building an expert system) (Neale, 1988).

Newell (1982) suggests that where possible the knowledge engineer should aim to eventually formulate a 'knowledge-level' description of the selected problem solving task; i.e. the types of knowledge necessary to solve that problem, including:

- the goals of the task;
- the actions of which the problem solver is capable; and
- the knowledge of which actions lead to which goals.

These are all connected by a principle of rationality, i.e. that 'if an agent has knowledge that one of its actions will lead to one of its goals, then the agent will select that action' (Newell, 1982, p. 102).

The purpose is to provide a clear specification of the minimal requirements for the expert system, preferably in the form of a conceptual model of expert problem solving, independent of the facilities of any expert system tool (see also Jackson, 1986b) and therefore distinct from the symbol level at which the knowledge is encoded. The function is to bridge the gap between verbal data on expertise and implementation, whilst avoiding the constraints implicit in the choice of any particular implementation formalism. This gap is commonly ignored in first generation expert systems, and the behaviour of the intelligent problem solver is displayed at the symbol level in the form of rules, frames, procedures etc. Modelling the behaviour instead at the knowledge level offers flexibility and predictability at a level above the confines of the representation.

An expert is rarely carrying out a single problem solving task in isolation in practice: a human expert may fulfil several functional roles (communication tasks or 'modalities') executing different problem solving tasks within each (see Basden, 1984, for discussion of roles). The distinction between modalities and problem solving tasks, noted by Breuker and Wielinga (1987a), has hardly been acknowledged. It is however important because expert systems sometimes cannot be switched from one modality to another without completely reconstructing the knowledge base. This was illustrated by Clancey in his attempts to convert MYCIN from a purely diagnostic system into a tutoring system (Clancey and Letsinger, 1981; Clancey, 1983; see also discussion in Coombs and Alty, 1984). The intended modality for the proposed expert system (e.g. advising, diagnosing, tutoring) should be decided before at an early stage therefore, so as to facilitate definition of the problem solving tasks within that modality for which conceptual models are required.

3.3.1 Machine-Aided Knowledge Acquisition

Gaines (1986) noted that the trend seems to be towards eliminating the knowledge engineer and getting the expert working directly with a shell. He predicted that knowledge acquisition tools will be a big development area in the next few years (Neale, 1988).

Several reasons to doubt that human labour is the appropriate solution for the 'knowledge engineering problem' have been identified by Shaw and Gaines (1987). These seem to reduce largely to economics (human labour as a proportion of expert systems development costs rising more rapidly than other costs), although Shaw and Gains also suggest that using a human intermediary may be less effective:

> 'knowledge may be lost through the intermediary and the expert's lack of knowledge of the technology may be less of a detriment than the knowledge engineer's lack of domain knowledge' (p 11).

Although many would argue against this – for example, Waterman, (1986, p. 154), advises succinctly 'don't be your own expert!' – the point here is that expert systems are not immune from universal trends towards packaging, pre-processing and convenience and away from hand crafting useful things from handcuffing useful things from raw materials. However, an important benefit *of research in tool development has been the attention focused on the issue of modelling expertise.*

The role of Machine Induction (MI) methods in knowledge acquisition is controversial. Proponents of MI have, for example, suggested that automated learning of decision rules is a solution to the knowledge acquisition bottleneck (e.g. Michie, 1983; Arbab and Michie, 1985; Michalski, 1983).

Machine (or Rule) Induction is a sub-field of Machine Learning. It requires the expert only to generate examples of a concept, or a classification problem rather than the expert and knowledge engineer jointly attempting to make explicit the expert's reasoning process by an arduous procedure of interviewing and analysis. The techniques themselves produce rules that reflect patterns in the examples provided, and the expert accepts or rejects patterns in the rules as being 'good' representations of his/her knowledge. Researchers in MI believe that this is an untapped and more effective use of the expert's knowledge, since they contend that experts find it much more straightforward to generate examples than to talk in abstract about solving problems (Martil, 1989; Martil, 1989, p. 49) further argues that:

> 'MI techniques provide the bare minimum to the knowledge engineer, the rules, and these are in a somewhat restricted form. The engineer must deal with any human engineering aspects, and design any support structures for explanations, object representation, etc.'

The current application of MI to the knowledge acquisition process is criticised by Martil (1989, p. 68) on two grounds, both of which stem from sound software development practice:

1. MI application techniques mix the design and implementation process;
2. The output from rule induction systems does not match the structure of the knowledge it is supposed to express.

There is however, a place for Machine Induction techniques as part of the whole knowledge elicitation process, but not as the sole techniques employed. They could be more appropriately employed after human labour oriented approaches have been invoked to do the groundwork, after establishing the context for instance (Graham, 1990).

3.3.2 Human Labour Oriented Approaches to Knowledge Elicitation

Johnson and Johnson (1987a) propose a methodology that uses a three phase elicitation process based around semi-structured interviews.

The first phase is to perform a broad, but shallow survey of the domain. This allows the elicitor to become oriented onto the domain, so a more flexible approach can be taken. This type of horizon broadening is a standard approach in social science research.

Once this shallow trawl of the domain has been done, the second phase requires that a more detailed task analysis is performed by the elicitor, focusing in on areas of interest. The structure of the interview uses a teachback technique to traverse the domain and validate elicitor understanding with the result that the elicitor progressively refines the model of the expert's competence. This model is qualitatively drawn up and uses a mediating representation, Systemic Grammar Networks. These are a context free, qualitative representation, which can be used as a tool for systems design, but their use does not imply the final use of any particular knowledge engineering shell or methodology.

The third phase of this approach is to validate the models drawn up from the expert with the wider expert community. The theoretical predictions of the model presented to the initial community used in the first phase, and then to a further independent population, to check the appropriateness and validity of the model which has been created.

This methodology has since been enhanced (Graham, 1990).

Domain orientation

The process of domain orientation is an important one, because if the elicitor focuses too quickly into the domain, then it is likely that any models or systems produced will be too narrow and inflexible to be of very much use. (The first phase of Johnson and Johnson's (1987a) methodology approximates to this process.)

The basic points for a knowledge engineer to consider in the domain definition phase have been summarised by Waterman (1996, pp. 129–132) and Neale (1988, p. 110). Prior to commencing work with the expert the knowledge engineer must be oriented with respect to the domain. This involves listing and investigating sources of information, establishing familiarity with basic vocabulary, identifying authoritative experts, and assessing the feasibility of the project.

Domain definition should result in the production of something like a Domain Definition Handbook (Grover, 1983), containing something of the following form:

- A general description of the problem;
- Bibliography of principal references;
- Glossary of terminology;
- Identification of experts;
- Characterisation of users;
- Definition of suitable measures of performance;
- Description of example reasoning scenarios.

Task analysis

P. Johnson *et al.* (1988), define task analysis as:

'. . . the investigation of what people do when they carry out tasks.'

An approach to task analysis involves a number of aspects, namely:

'. . . how the data are collected; a description of that data; a method of analysing tasks and; a representational framework for modelling those tasks.'

The task analysis stage of knowledge elicitation is an important one. A variety of regimes for this stage have been presented in the literature, although most are still informal and unproven (Wilson, 1989).

L. and N. Johnson's (1987a) approach to this has already been discussed. Wilson *et al.* (1986) gives some other approaches to it.

Different analysis techniques may focus on different aspects of a task, but they all use the same basic observational techniques. The issues associated with systematic observation are non trivial (e.g. Diaper, 1987a).

The purpose of task analyses (which themselves can be treated as knowledge elicitation methods, and this is why they are often not listed separately) is principally to establish the actual requirements of users and to formally identify the possible range of tasks and sub-tasks that the expert system or knowledge-based system may be expected to perform (Diaper, 1989).

Other approaches to task analysis also include Clancey's heuristic classification (Clancey, 1985), and Chandrasekaran et al's Generic Task approach (Chandrasekaran (1983; 1986; 1987)). Clancey (1989, p. 11) defines the former approach as:

'. . . *a problem solving method with which programs select solutions from predescribed alternative.*'

The latter approach (Chandrasekaran, 1989, p. 183) proposes that:

'. . . *knowledge systems should be built out of basic building blocks, each of which is appropriate for a basic type of problem solving.*'

3.4 KNOWLEDGE ELICITATION TECHNIQUES

The statement, 'that the problem of knowledge acquisition is the critical bottleneck' (Feigenbaum and McCorduck, 1983), has engendered frustration in knowledge engineers that has contributed to the misguided belief of some, that there is 'one magical technique' (Kidd, 1987). It is more helpful to develop a 'tool kit of techniques', from which to select according to the nature of the domain and the type of knowledge under investigation (Gammack and Young, 1985; Berry and Broadbent, 1986; Banathy, 1988); remembering that 'nothing is so detrimental to the acquisition of the complexity of the knowledge and reasoning in a particular domain' (Breuker and Wielinga, 1983, p. 6).

Knowledge elicitation techniques may be broken down into several types (Neale, 1988):

- psychological techniques;
- observation;
- multidimensional techniques; and
- protocol analysis.

Neale describes in total twenty-four techniques for knowledge elicitation within these categories. Johnson *et al.* (1988) list a further ten techniques and Cordingly (1989), notes another six, in addition to those described by Neale. Figure 3.2 presents 'A Reference Table of Techniques for Knowledge Elicitation' (Graham, 1990), as described/listed by Neale, Johnson *et al.* and Cordingly combined. Sub-dividing these into:

- psychological techniques;
- observation;
- multi-dimensional techniques;
- protocol analysis, as imputed by Neale; and
- 'other'.

TECHNIQUES					
CATEGORY	Pyschological techniques	Observation	Multidimensional techniques	Protocol analysis	Other
	Tutorial interview	Dialogues (non participant)	Card sorting	Concurrent protocols	Critiquing
	Focussed interview	Participant observation	Multidimensional scaling (MDS)	Retrospective protocols	Role play
	Distinction of goals		Repertory grid		Simulation
	Reclassification		Frequency counts		Object tracing
	Dividing the domain		Proximity analysis		'Listening'
	Systematic system -to-fault		Matrix techniques		'Collection' of artefacts
	Intermediate reasoning steps				Knowledge competitions
	Structured interview and questionnaires				Group discussions
	Twenty questions				Multi-choice questions
	Laddered grid				Analyst to carry out task with instruction
	Teachback interview				Observation with knowledgeable commentator
	Ethnographic interview				Sample outputs
	Introspection				
	Retrospective case study				
	Critical incident				
	Forward scenario simulation				

FIGURE 3.2.   A Reference Table of Techniques for Knowledge Elicitation

This table may provide a useful guide to the knowledge engineer when considering what available techniques should be applied to his/her particular domain of application. Full accounts of these techniques may be found in Neale (1988), Johnson *et al.* (1988) and Cordingly (1989) respectively. The table given is in no way a complete list of all available techniques, it is merely a selection of the more familiar ones.

3.5 KNOWLEDGE ELICITATION ISSUES

It is suggested by the approach of some enthusiastic promoters of expert systems, that all you need in order to make a computer program intelligent is to provide some general rules and lots of very specific knowledge. A mechanistic conception implicit in the very name 'knowledge engineering' coined to describe the activity of building expert systems by Feigenbaum (Buchman and Duda, 1983, give an early description of the task), has thus been created (Neale, 1988).

There are two main approaches to the role of the elicitor in the elicitation process. The first views knowledge elicitation as a static process, involving the elicitor taking an authoritarian role in telling the expert what the elicitor requires, with the elicitor just picking up the pearls that the elicitor has scattered (Blackler, 1989). This view is widely held in the field of knowledge engineering. Feigenbaum and McCorduck (1983) describe the job of the knowledge engineer as to:

'mine those jewels of knowledge out of (experts') heads one by one.'

This type of approach is not only methodologically unsound but, given the experts' high position within their chosen domain, it is very likely to cause any of them to become alienated and disaffected with the entire process of elicitation (Neale, 1988).

An alternative approach looks at knowledge not as a substance to be quarried (Young, 1988), but rather as something to be probed by the knowledge engineer as an essentially social process (Goodall, 1985). The aim of such a process would be to model the expert's knowledge rather than just extracting data.

If, as in the case of the first approach, the model is just considered as a static collection of knowledge which can be transferred from an expert to an expert system, then this process is simply poor systems design. It will often lead to undesirable systems which date rapidly. (XCON (Bachant and McDermott, 1984; McDermott, 1981) is an example of this).

The second is a more useful approach to elicitation, which has the elicitor and the expert both playing an interactive role in a negotiation cycle. Fewer psychological and practical difficulties will be encountered if knowledge acquisition is considered as social science research, aiming to model the expert's knowledge rather than merely extracting data (Breuker *et al.*, 1987; Johnson and Johnson, 1987b; Kidd, 1987). Additionally with Evans (1987) it is argued that the concepts and methods of cognitive psychology have often been unjustifiably ignored in the elicitation of human knowledge.

This type of model was used by Hawkins (1983) to elicit data for mineral exploration expert systems; the essential features of the negotiation cycle are:

1. The elicitor first elicits data about the problem from the expert;
2. He/she develops a 'minimal model' that accounts for the data provided by the expert;
3. He/she generates advice based on the model, and feeds this back to the expert;
4. The expert may accept or query the advice, and, possibly therefore, accept or query the model;
5. The queries lead to further data elicitation, and the repeat of the elicitation/modelling/query cycle.

In this model of the elicitation process both the expert and the elicitor play an active role in constructing and evaluating the models of the expert's competence. Although the expert is part of the negotiation cycle, the central role is that of the elicitor who must elicit data from the expert, use the data to form a model and apply that model, in the form of advice, which will be discussed with the expert with the view to improving the model. So the role of the elicitor is central and active, rather than the passive and peripheral elicitor of Feigenbaum and others (Blackler, 1989).

Gaines (1989) also encourages the seeing of the expert as part of an expert community which, via an implicit system of rewards and criticism, has helped the expert develop to his current level, and has also produced a body of public knowledge which includes, theories, case histories and strategies that can be used to supplement the data from the expert at the modelling, data collection and advice generation phases of the negotiation cycle. Gaines emphasises that such data should be used to supplement, not replace expert testimony. Thus the negotiation cycle is a method of progressively refining the competence model, keeping both the expert and the elicitor involved in the process. This does however, imply a great deal of work on the part of the elicitor.

Johnson and Johnson (1987a) use a very similar type of approach which uses a three phase elicitation process based around semi-structured interviews. (This approach was discussed in more detail in section 3.3.2.)

3.6 KNOWLEDGE ACQUISITION TOOLS: TEIRESIAS AND OPAL

TEIRESIAS (Davis, 1980b), was part of EMYCIN's interface with the system designer. The TEIRESIAS program was a 'knowledge editor' devised to help with the development and maintenance of large knowledge bases. The program concentrated on the syntax of the production rules in an evolving expert system, making sure that new rules referenced medical parameters referenced in similar, extant rules.

TEIRESIAS possessed no knowledge of either the domain of application or the problem-solving strategy to be employed by the system under construction. This manifested as both a strength and a weakness. Generality; such a syntactic analysis can be applied to rules in almost any domain, was the strength of the method. The weakness lies in the fact that it places a considerable burden on both the expert and the knowledge engineer; they are the sole repositories of all the background knowledge about the domain on which the decision rules are based. In other words, they alone are responsible for ensuring that the rules make sense at any level deeper than that of syntactic conformity. Nevertheless, TEIRESIAS contained many innovations.

In contrast, OPAL (Musen *et al.*, 1987), attempts to provide acquisition strategies which are guided by knowledge of the domain. Much of the detail of how knowledge is represented and deployed is concealed from the user. OPAL sets out to elicit knowledge from an expert directly by an 'interview' session conducted at the terminal. OPAL is not a general purpose program, however; it uses knowledge about a particular domain of application (cancer therapy) to elicit treatment plans from which decision rules can be generated.

OPAL expedites knowledge elicitation for the expert system ONCOCIN (Shortliffe *et al.*, 1981), which constructs treatment plans for cancer patients. It uses a domain model to acquire knowledge from an expert via a graphical interface. It also uses domain knowledge to translate the information acquired at the terminal into executable code, such as production rules and finite state tables.

Musen *et al.* (1987) note that knowledge acquisition posed considerable problems in developing the original prototype of ONCOCIN. The rationale for developing OPAL was to speed up the acquisition process by reducing dependence on the knowledge engineer as the transcriber and translator of expertise.

OPAL's form-filling approach is effective, because of the incorporation of domain assumptions. Outlining these assumptions itself involves a knowledge engineering effort. Once that investment has been made, subsequent knowledge elicitation is greatly facilitated. OPAL's success illustrates the benefit of viewing domain knowledge at different levels of abstraction instead of focusing solely on implementation details.

A feature of many new acquisition systems is the use of this technique of eliciting domain knowledge from an expert by an interview conducted at the terminal. ETS (Boose, 1986) and Student (Gale, 1986) use something analogous to form-filling to read information into structured objects like frames. Not all such systems however, have the graphical sophistication of OPAL, and not all compile such knowledge directly into decision rules. Knowledge elicitation in OPAL is made notably easier by the highly structured and stylised nature of cancer therapy plans. The lessons learnt are that:

- Knowledge acquisition is greatly facilitated by being itself knowledge based; and
- Knowledge elicitation is a substantial problem in itself.

The knowledge that one needs in order to acquire more knowledge can be viewed as a form of meta knowledge. It is mostly about structure and strategy, involving information about ways of classifying phenomena (such as diseases) and ways of deciding between alternative courses of action (such as therapies). This is also the kind of knowledge needed to explain solutions.

Automated approaches to acquisition are not restricted to knowledge elicitation by interview based on a domain model. Other approaches include: acquisition strategies organised around a particular problem solving method, and machine learning from examples.

4 Knowledge Representation

D. GRAHAM

4.1 SCHEMES FOR REPRESENTING KNOWLEDGE

A representation is a convention about how to describe things, like a language. A representation makes some aspects of the entity it describes explicit and leaves others implicit. For example, Vision: gray scale array makes brightness explicit, primal sketch makes brightness changes explicit, 21/2 D sketch makes surfaces explicit, 3D models makes objects explicit (Marr, 1982).

It is useful to distinguish between a representational *scheme* and the *medium* of its implementation (Hayes, 1974). Categories of representational schemes (Mylopoulos and Levesque, 1984):

- *Logical representation schemes* – This class of representation uses expressions in formal logic to represent a knowledge base. Inference rules and proof procedures apply this knowledge to problem instances. First-order predicate calculus is the most widely used logical representation scheme, but it is only one of a number of logic representations (Turner, 1984). PROLOG is an ideal programming language for implementing logical representation schemes;
- *Procedural representation schemes* – Procedural schemes represent knowledge as a set of instructions for solving a problem. This contrasts with the declarative representations provided by logic and semantic networks. In a rule-based system for example, an if . . . then rule may be interpreted as a procedure for solving a goal in a problem domain: to solve the conclusion, solve the premises in order. Production systems are examples of a procedural representation scheme;
- *Network representation schemes* – Network representations capture knowledge as a graph in which the nodes represent objects or concepts in the problem domain and the arcs represent relations or associations between them. Examples of network representations include *semantic networks* , *conceptual dependencies*, and *conceptual graphs;*

41

- *Structured representation schemes* – Structured representation languages extend networks by allowing each node to be a complex data structure consisting of named slots with attached values. These values may be simple numeric or symbolic data, pointers to other frames, or even procedures for performing a particular task. Examples of structure representations include *scripts*, *frames*, and *objects*.

4.2 ISSUES IN KNOWLEDGE REPRESENTATION

Much of the research work in knowledge representation is motivated by the effort to program a computer to understand human languages such as English, because of the breadth and complexity of the knowledge required to understand natural languages. A number of important issues have dominated this research (Luger and Stubblefield, 1989).

It is not sufficient to examine the semantics of a natural language and implementation structures that seem to capture these features, but more attention should be paid to the semantics of the representation formalism itself (Woods, 1975).

The representation language is the basis for all inferences that will be made by the system and therefore determines what can be known and ultimately expressed. There are a number of considerations in designing a representation language:

1. Exactly what 'things' can be represented by objects and relations in the language?
2. Granularity of the representation;
3. How can representations distinguish between *intensional* and *extensional* knowledge?
4. How can meta-knowledge best be represented?
5. Inheritance.

We cannot describe knowledge representation in great detail, so we will now constrain our reporting to knowledge representation schemes commonly used by knowledge-based systems.

4.3 LOGICAL REPRESENTATION SCHEMES

Logical representations grew out of the efforts of philosophers and mathematicians to characterise the principles of correct reasoning. The major concern of this work is the development of formal representation languages with sound and complete inference rules. Consequently, the semantics of predicate calculus emphasise *truth-preserving* operations on well-formed expressions.

An alternative line of research has grown out of the efforts of psychologists and linguists to characterise the nature of human understanding. This work is concerned less with establishing a science of correct reasoning than with describing the way in which humans actually acquire and use knowledge in their world. As a result, this work has proved particularly useful to the AI application areas of natural language understanding and common-sense reasoning. Problems arise in mapping common-sense reasoning onto formal logic.

> Example: The operators **V** and => are commonly thought of as corresponding to the English 'or' and 'if . . . then'. However, these operators are concerned solely with truth values and ignore the fact that the English 'if . . . then' generally suggests a specific relationship (generally causal) between its premises and its conclusion.

For example, the English sentence 'If the snowman is a cardinal, then it is white' may be written in predicate calculus as:

VX(cardinal(X)=>white(X))

This may be changed, through a series of truth-preserving operations, into the logically equivalent expression:

VX($\neg$ white(X)=> $\neg$ cardinal(X))

These two expressions are logically equivalent; that is, the second is true if and only if the first is true.

While logically sound, this line of reasoning strikes as meaningless and rather silly. Truth value equivalence overlooks the more subtle connotations of the original English sentence. The reason is, that the logical implication only expresses a relationship between the truth values of its operands, while the original English sentence implies a causal relationship between membership in a class and the possession of properties of that class.

The limitations of predicate calculus are largely a result of Tarskian semantics; the assignment of a truth value to logical expressions based on an interpretation in some possible world. It is not always enough to know that a statement such as 'snow is white' is true; the meaning of the atomic symbols must also be addressed (Woods, 1975).

The representation should be able to answer questions about these objects such as: 'What is snow made of?' 'What temperature is Frosty the Snowman?' This requires that the representation not only preserves truth values when combining expressions but also has some means of making truth assignments based on knowledge of a *particular* world.

4.4 PROCEDURAL REPRESENTATION SCHEMES: PRODUCTION SYSTEMS

Production systems employ *production rules*, sometimes called *condition-action rules* or *situation-action rules*. Production rules previously used to some extent in automata theory, formal grammars and the design of programming languages, are now being exploited for psychological modelling (Newell and Simon, 1972) and expert systems (Buchnan and Feigenbaum, 1978).

Production rules function as generative rules of behaviour; given some set of inputs, the rules determine what the output should be. Just as the functions of LISP or the relations of PROLOG do, productions define a programming paradigm.

In expert systems, production rules are more prescriptive (imperative) rather that descriptive as in LISP and PROLOG.

Canonical systems (Post, 1943) are formal systems based on: an alphabet for making strings; some set of strings that are taken as axioms; and a set of productions, composed of an antecedent and consequent. Canonical systems appear trivial, rewriting one string of symbols into another, but all calculi of logic and mathematics are just sets of rules to manipulate symbols.

In problem solving, as in expert systems, we are interested in taking a representation of some problem and transforming it until it satisfies some criteria.

The alphabet of canonical systems is replaced by a vocabulary of symbols or atoms, and a rather simple grammar for forming symbol structures. The vocabulary typically consists of three sets: a set of names of objects in the domain; a set of property names that impute attributes to objects; and a set of values that these attributes can take. Names and values can overlap. The grammar used for example is that of object-attribute-value triples, for example:

(Frosty scarf-colour red)

Numbers of triples for some object are often combined into a vector of the form:

(Frosty ∧scarf-colour red ∧scarf-fabric wool)

On attaining a vocabulary of symbols and a grammar for generating symbol structures, the initial state of some problem of interest can be encoded. This representation corresponds to the axioms of a canonical system; these are the symbol structures that are going to be progressively rewritten in a series of rule applications.

A *production system* consists of a *rule set* (occasionally called *production memory*), a *rule interpreter*, and a *working memory*. The rule interpreter decides when to apply which rules. The working memory holds the data, goal statements and intermediate results that make up the current state of the problem; it is the central data structure which is examined and modified by productions. Rules are triggered by this data, and the rule interpreter controls the activation and selection of rules at each cycle.

Production system rules have the general form:

if *premises* P1 and . . . Pm are true
then perform *actions* Q1 and . . . Qn

Premises are sometimes referred to as 'conditions' (or the left-hand side of the rule), actions 'conclusions' (or the right-hand side of the rule).

Premises are normally represented by object-attribute-value vectors. A very simple example:

(Snowman ∧Name Frosty ∧Purpose Play ∧Season Winter)

A rule including this condition:

(p Toy)
    (Snowman ∧Name Frosty ∧Purpose Play ∧Season Winter)
       -> (MAKE ∧Toy-list ∧Add Frosty)

Toy rule signifies that, if Frosty's purpose is play, Frosty is a toy. The interpreter's role is:

1. To *match* premise patterns of rules against elements in working memory;
2. Apply *conflict resolution* if there is more than one rule that could fire;
3. *Apply* the rule, possibly adding to, or deleting from working memory. Go to (1).

Premises are patterns that are meant to match vectors in working memory. Actions, like MAKE, adds the new vector (∧Toy-list ∧Add Frosty) to working memory.

Global and *local* control are two general approaches to controlling the behaviour of rule-based systems. Control is not a trivial problem. Global control regimes tend to be domain-independent 'hard-coded' (i.e. difficult for programmers to change), local control regimes tend to be domain-dependent and 'soft-coded'. The former strategy does not use domain knowledge to any significant degree. The latter uses special rules, called *meta rules*, because they reason about which (object-level) rule to fire. These require domain knowledge to reason about control.

Commonly used conflict resolution mechanisms are *refractoriness*, *recency* and *specificity*. Such strategies are often used in combination to form a global control regime. At the global level of control, production rules can be driven forward or backward:

- Forward chaining: chain forward from conditions that are known to be true towards problem states which those conditions establish;

- Backward chaining: chain backward from a goal state towards the conditions to be established.

This is like the forward and backward strategies in resolution theorem proving. Forward chaining is often analogous to 'bottom-up' reasoning (from facts to goals), backward chaining analogous to 'top-down' reasoning from goals to facts.

An *AND/OR tree* is a useful device for representing the search space associated with a set of production rules. Rule-based programming does not solve the combinatorial explosion, because the AND/OR tree for any problem may branch exponentially.

4.5 NETWORK REPRESENTATIONS: SEMANTIC NETWORKS

Much of the research in network representation has been done in the arena of natural language understanding. In the general case, natural language understanding requires far more knowledge than the specialised domains of expert systems. It includes an understanding of common sense, the ways in which physical objects behave, and how the interactions which occur between human institutions are organised. A natural language program must understand intentions, beliefs, hypothetical reasoning, plans, and goals. Because it requires such large amounts of broad-based knowledge, natural language understanding has always been a driving force for research in knowledge representation.

Knowledge is generally structured in terms of specific relationships such as object/property, class/subclass, and agent/verb/object. Research in network representations has often focused on the specification of these relationships.

Early work relating to *Associationist* theories defined the meaning of an object in terms of a network of associations with other objects in a mind or knowledge base. Although symbols denote objects in a world, this denotation is mediated by our store of knowledge. There is psychological evidence that in addition to their ability to associate concepts, humans also organise their knowledge hierarchically, with information kept at the appropriate levels of the taxonomy. Collins and Quillian (1969) modelled human information storage and management using a semantic network. (Vis Semantic network developed by Collins and Quillian in their research on human information storage and response time (Harmon and King, 1985).

Graphs, by providing a means of explicitly representing relations using arcs and nodes, have proved to be an ideal vehicle for formalising associationist theories of knowledge.

A semantic network represents knowledge as a graph, with the nodes corresponding to facts or concepts and the arcs to relations or

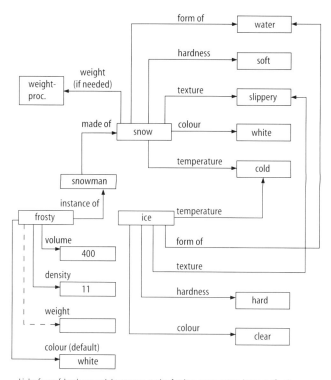

Links: form of, hardness, weight, texture, made of, colour, temperature, instance of, volume, density.
Facts/Concepts: water, soft, weight-proc, slippery, snow, white, cold, snowman, frosty, 400, 11, ice, hard, clear.

FIGURE 4.1   Network Representation of the Properties of Frosty

associations between concepts. Both nodes and links are generally labelled.

The term semantic network encompasses a family of graph-based representations. These representations differ chiefly in the names that are allowed for nodes and links and the inferences that may be performed on these structures. However, a common set of assumptions and concerns is shared by all network representation languages. Conceptual graphs (Sowa, 1984) are a modern network representation language that integrates many of these ideas.

Summary of basic features

- Semantic nets (Winston, 1992) denote facts or concepts and describe relations or associations between them;
- Facts or concepts are denoted by labelled *nodes*. Relations or associations are denoted by labelled arrows (*links*);
- A node-and-link net is a semantic net only if there are semantics!

Intuitive meaning: Example re. Figure 4.1, *frosty instance of snowman made of snow, form of water*, etc.

The semantic network employs natural-language labels which associate intuitive meanings to nodes and links, thereby producing informal semantics. Many people regard such semantics as slip-shod or dangerous. Informal intuition-based semantics is imprecise in the extreme. The only semantic statement made is that nets are related to objects, actions, and events by the vague, individual, ambiguous, and time varying intuition of the beholder.

Intuition-based semantics is rejected in favour of: either

- Equivalence semantics;
- Procedural semantics; or
- Descriptive semantics.

Inheritance, Demons, Defaults and Perspectives

Links impinge on *value* nodes. *Slot*s of a node correspond to the different named links.

Inheritance

Several benefits come from inheritance systems. They allow us to store information at the highest level of abstraction, which reduces the size of knowledge bases and helps prevent update inconsistencies.

Inheritance enables descriptions of movement from classes to instances, and makes class membership and subclass relations explicit. Properties of instances are obtained from classes by inheritance through *instance of; made of; form of;* links. To obtain results corresponding to our expectations, we need a search procedure that looks for appropriate slots in appropriate nodes (a general requirement).

Inheritance is a relation by which an individual assumes the properties of its class and by which properties of a class are passed on to its subclass. Inheritance is used to implement *default values* and *exceptions*.

Type hierarchies may assume a number of different forms, including *trees, lattices,* or *arbitrary graphs.* In a tree hierarchy, each type has only one supertype. This is a simple and well-behaved model, but it is not as expressive as hierarchies that allow multiple types or let types have multiple supertypes.

Although multiple inheritance hierarchies can introduce difficulties in the definition of representation languages, their benefits are great enough to offset these disadvantages. The lattice is a common form of multiple inheritance.

It is often helpful to attach procedures to object descriptions. These procedures, sometimes referred to as *demons*, are executed when an object is changed or created and provide a vehicle for implementing graphics I/O, consistency checks, and interactions between objects.

Demons

Enable access to initiate action. Makes a procedure's purpose explicit, and relates it to the object class it is relevant to. When we have no values, we can compute a value using existing information. The procedures needed for such computations here are *if needed procedures,* also called *if needed demons*. If needed procedures can be inherited (eg weight in figure 4.1).

Defaults

Enable assumptions in lieu of fact. Makes unknown but likely values explicit. The ordinary value type i.e. the one corresponding to certain truth, is said to be in the *value* facet of a slot. The default value type associated with probable truth is said to be the *default* facet.

In the absence of specific information defaults establish probable value, usually through inheritance (*form of, made of, instance of* links). Values in facets other than the *value* facet are identified here by the appearance of parentheses. Example, *colour* of frosty (figure 4.1).

Inheritance procedures

N inheritance so called because the search path looks like an N. Another kind of inheritance is called Z inheritance, because the search path looks like a Z.

Perspectives

Enable purpose to guide access. Makes context-sensitivity explicit. Nodes broken-up into node bundles in which each bundle member specifies one named or unnamed perspective. If we ask for a value without specifying a perspective, the node bundle is treated as if it were collapsed into a single node. Answers, however, are annotated to include the name of the perspective from which they come, if any.

Inheritance for objects which have parts

We can pass values from snowman to frosty, and from parts of frosty to snowman. These are virtual nodes and links – they need only be created if we need to 'hang' something from them (i.e. attach some value, such

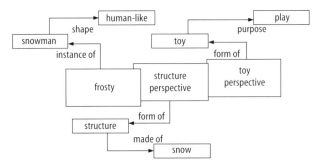

Objects may be viewed from various perspectives. Here frosty is described by a node bundle in which one element is an unnamed general perspective (frosty), another corresponds to frosty viewed as a structure, and another to frosty viewed as a toy.

FIGURE 4.2.   Viewing Objects from Perspectives

as the colour or part). But we can only do this if it is clear what parts correspond to what, as matching ambiguity stalls inheritance.

Standardisation of Network Relationships

In itself, a graph notation of relationships has little advantage over predicate calculus; it is just another notation for relationships between objects. The power of the network representation comes from the definition of links and associated inference rules that define a specific inference such as inheritance.

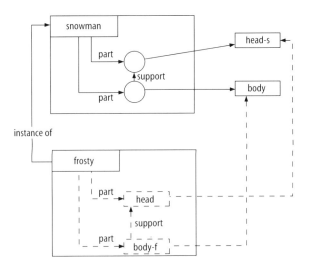

FIGURE 4.3   Inheritance for Objects which have Parts

4.6 STRUCTURED REPRESENTATIONS

4.6.1 Conceptual Dependency Theory

A number of major research endeavours attempted to standardise link names further (Masterman, 1961; Wilks, 1972; Schank and Colby, 1973; Schank and Nash-Webber, 1975). Each effort worked to establish a complete set of primitives that could be used to represent the deep structure of natural language expressions in a uniform fashion. These were intended to assist in reasoning with language constructs and to be independent of the idiosyncrasies of individual languages or phrasing.

Conceptual dependency theory (Schank and Rieger, 1974), offers a set of four equal and independent primitive conceptualisations from which the world of meaning is built, namely:

- ACTs   actions;
- PPs   objects (picture producers);
- AAs   modifiers of actions (action aiders);
- PAs   modifiers of objects (picture aiders).

All actions are assumed to reduce to one or more primitive ACTs for example. These primitives, listed below, are taken as the basic components of action, with more specific verbs being formed through their modification and combination:

- ATRANS   transfer a relationship (give);
- PTRANS   transfer a physical location of an object (go);
- PROPEL   apply physical force to an object (push);
- MOVE   move body part by owner (kick);
- GRASP   grab an object by an actor (grasp);
- INGEST   ingest an object by an animal (eat);
- EXPEL   expel from an animal's body (cry);
- MTRANS   transfer mental information (tell);
- MBUILD   mentally make new information (decide);
- CONC   conceptualise or think about an idea (think);
- SPEAK   produce sound (say);
- ATTEND   focus sense organ (listen).

These primitives are used to define *conceptual dependency relationships* that describe meaning structures as case relations or the association of objects and values.

Conceptual dependency relationships are *conceptual syntax rules* that constitute a grammar of meaningful semantic relationships. These relationships can be used to construct an internal representation of an English sentence. Schank supplies a list of attachments or modifiers to the relationships to add tense and node information to the conceptua-lisations. These relations are the first-level constructs of the theory, the simplest semantic relationships out of which more complex structures can be built.

Complex structures can be built with conceptual dependencies, for example, the conceptual dependency representation of the sentence 'John prevented Mary from giving the book to Bill' (Schank, 1974). This example demonstrates how causality may be represented. Conceptual dependency theory gives a number of important benefits. Problems of ambiguity are reduced by providing a formal theory of natural language semantics. The representation itself also directly captures much of natural language semantics.

Conceptual dependency attempts to provide a *canonical form* for the meaning of sentences. *That is, all sentences that have the same meaning will be represented internally by syntactically identical, not just semantically equivalent, graphs.* This is an effort to simplify the inferences required for understanding.

It is questionable whether a program may be written to reliably reduce sentences to a canonical form. Woods (1975) and others have pointed out, that reduction to canonical form is provably uncomputable for monoids, a type of algebraic group that is far simpler than natural language. Also, there is no evidence that humans store their knowledge in any sort of canonical form.

Other criticisms include the computational price paid in reducing everything to such low-level primitives, and that the primitives are not adequate to capture many of the more subtle concepts that are important in natural language.

Conceptual dependency relationships are used to represent the basic components of *scripts* (described in section 4.6.3).

4.6.2 Frames

When one encounters a new situation (or makes a substantial change in one's view of a problem) one selects from memory a structure called a 'frame'. This is a remembered framework to be adapted to fit reality by changing details as necessary (Minsky, 1975).

Frames, as well as *object-oriented systems*, provide a vehicle for an organisation, representing knowledge as structured objects consisting of named slots with attached values. The notion of a frame or schema as a single complex entity is thus reinforced by this notion.

The slots in the frame contain information such as:

1. Frame identification information;
2. Relationship of this frame to other frames;
3. Descriptors of requirements for frame match;

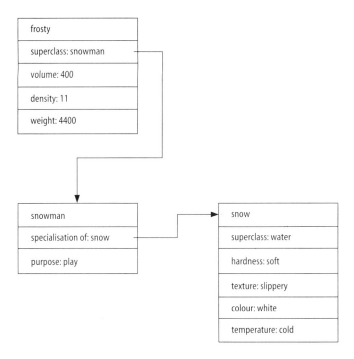

FIGURE 4.4   Proportion of the Frame Description of Frosty

4. Procedural information on use of the structure described. This ability to attach procedural code to a slot, is an important feature of frames;
5. Frame default information;
6. New instance information.

Frames add to the power of semantic nets by allowing complex objects to be represented as a single frame, rather than a large network structure, thus also facilitating the representation of stereotypic entities, classed, inheritance, and default values. Frames, like logical and network representations, are a powerful tool, however, many of the problems of acquiring and organising a complicated knowledge base must still be solved by the programmer's skill and intuition.

Because of their power and generality, frames have nevertheless emerged as an important representation in AI.

4.6.3 Scripts

A *script* (Schank and Abelson, 1977) is a structured representation describing a stereotyped sequence of events in a particular context. The most famous example of such is probably the restaurant script (Schank, 1977).

The components of a script are:

1. *Entry conditions* or descriptors of the world that must be true for the script to be called;
2. *Results* or facts that are true once the script has terminated;
3. *Props* or the 'things' that make up the content of the script. This allows reasonable default assumptions about the situation;
4. *Roles* are the actions that the individual participants perform;
5. *Scenes* Schank breaks the script into a sequence of scenes each of which presents a temporal aspect of the script.

The elements of the script, the basic 'pieces' of semantic meaning, are represented using conceptual dependency relationships. Placed together in a framelike structure, they represent a sequence of meanings, or an event sequence. Script *roles* help resolve pronoun references and other ambiguities. Scripts can also be used to interpret unexpected results or breaks in the scriptal activity.

Scripts, like frames and other structured representations are subject to certain problems, including the *script match* problem and *between-the-lines* problem. The script match problem is 'deep' in the sense that no algorithm exists for guaranteeing correct choices. It requires heuristic knowledge about the organisation of the world, and scripts assist only in the organisation of that knowledge. The between-the-lines problem is equally difficult. It is not possible to know ahead of time the possible occurrences that can break a script. Reasoning can be locked into a single script, even though this may not be appropriate. These problems are not unique to script technology but are inherent in the problem of modelling semantic meaning. Eugene Charniak (1972) has illustrated the amount of knowledge required to illustrate even simple children's stories.

Type Hierarchies, Inheritance, and Exception Handling

Inheritance across class or type hierarchies is an important feature of almost all network and structured representation languages. Inheritance is implemented through relations such as the type hierarchy of conceptual graphs, as well as through various links including ISA, AKO (A Kind Of), and SUPERC (SUPER Class).

A number of anomalies arise in inheritance (Brachman, 1985a; Touretzky, 1986), particularly in multiple inheritance hierarchies and in the use of inheritance to implement defaults and exceptions.

4.6.4 Objects

Like frames, objects are organised with data and procedures together related by some inheritance mechanism. Objects differ from other previously mentioned knowledge representation schemes, by their ability to inherit and combine procedures as well as data. They also possess a special message-passing protocol to enable communication between objects. COMMON LISP Object System (CLOS) (Keene, 1989), is an example of an object-oriented system.

Object-oriented programming techniques, with their interacting objects and agents, provide a framework for many classes of problem, especially those which possess a strong simulation component e.g. scheduling.

Consideration of the kinds of objects and behaviour relevant to the problem by AI programmers, is encouraged by the techniques of procedure and data abstraction necessitated by this approach.

5 Neural Networks

D. MARTLAND

5.1 INTRODUCTION

The study of neural networks has become fashionable in recent years, but dates back to before the start of this century, when Golgi and Ramon y Cajal were carrying out detailed studies of biological neurons.

More recent work has been on the development of systems based on artificial neurons, and practical applications, but the origins of the work are rooted in the understanding of biological systems. This chapter introduces the biological and historical background, and then describes a number of artificial neural models used for a wide variety of applications. Consideration is also given to some of the problems faced by the use of artificial models in practical situations.

5.2 BACKGROUND TO NEURAL NETWORKS

In the nineteenth century, theories of neuron structure and behaviour were primitive, and the concept of neuron itself was not accepted until Ramon y Cajal showed that neurons formed distinct entities, by the use of staining techniques developed by Golgi. Before that, it had been thought that nervous systems formed an integrated whole, and individual neurons were not recognised. Even after this, it was not known how neurons operated, and there were several theories, including one which suggested that neurons functioned as electrical components. McCullogh and Pitts (1943) reported studies of individual neurons, which demonstrated their electrical behaviour, including the so-called all-or-none output behaviour, in which a neuron fires if stimulated above a set threshold level, but exhibits a passive response otherwise.

Analysis of the interaction of such firing neurons showed that the transmission delay between individual neurons was too large to be explained by electrical transmission, being of the order of one millisecond, and thus chemical mechanisms were proposed. Modern theories account for the behaviour of chemical neurons in terms of release of neuro-transmitters in a controlled fashion. For a detailed description of the action of neuro-transmitters in such neurons, see Alberts *et al.* (1983).

Although most biological neurons are assumed to transmit information by chemical means, there is evidence for electrical neurons in specialised circuits, for example, in some fish.

In large natural systems neurons are used both within the brain and spinal chord, which form part of the *central nervous system* (CNS) and the *peripheral nervous system* (PNS). These include *receptors* such as the retinal rods and cones in the eye, the ciliary cells in the ear and vestibular system, taste buds and other *sensory neurons* and *motor neurons* which are used to activate muscles. Some neurons may simply be used to link other neurons together, and these would be classified as *interneurons*. In humans there are thought to be about 10^{10} neurons in the brain and a significant proportion of these, perhaps 60 per cent are thought to be connected with vision. Each neuron may connect with another 10,000 neurons, which suggests that the total number of network connections could be around 10^{14}.

5.2.1 Motivation

There are many motivations for studying neural networks, but undoubtedly a significant motivation is that so far artificial computer systems have not succeeded in tasks which require learning of sophisticated operations in real time. Thus, robot systems can be programmed to perform manufacturing operations by following preset instructions, but they will do so only under program control. A human can be taught to perform such operations by being shown the sequence of steps to be performed, but should then be able to vary the operation sequence to cope with variations in the environment.

Pursuing this analogy, it might be possible to devise neural networks that can learn from their experience and are also robust to failure of individual neurons, or incomplete input, in ways similar to those used by animals and humans. For many workers, it is these aspects of artificial neural modelling that are of greatest interest.

However, it must be said that there are other views. Some workers are specifically interested in the behaviour of particular natural neural networks, for example, to study the behaviour of humans with specific cerebral deficits. Others will be interested in studying the behaviour of individual neurons and may be motivated into helping sufferers of certain kinds of neurological disorders.

Workers in pattern recognition and image processing will be particularly interested in the following possibilities for applying artificial neural networks:

- parallel operation – neural networks can be implemented on parallel hardware systems, which may give a substantial speed up in processing;
- sophisticated filtering – neural networks can be constructed for sophisticated filters, either by prescriptive procedures, or by a training process. For applications such as cell screening where many images may need to be scanned, this could be very effective in saving human effort and may be more reliable;
- decision making – neural networks do not only have to operate on image data, but can also be trained to make or recommend decisions in many practical applications. Unlike rule-based knowledge-based systems, such networks usually learn by example, although there may be significant advantages in incorporating known rules into a neural network, or a hybrid decision making system-based on both ruled-based KBS and neural network components;
- data compression and dimensionality reduction – techniques for compressing data based on neural network methods have been devised and demonstrated. These could be useful where large images are to be transmitted over a network, although they are not so likely to be used for storage, since the data compression is lossy. Other methods, such as fractal data compression or block image compression based on the use of discrete cosine transforms, are likely to be more effective.
- image classification – for libraries of images, which will become more important in future years, there is likely to be a need for rapid retrieval of images with specific features. This may require pre-processing to identify characteristics for tagging each image, or may be done on the fly, when the request for images is made. Neural network algorithms may be useful in either locating the required images, or reducing the size of the image data base which needs to be examined.

5.2.2 Biological Neurons

The structure of a typical chemical neuron is as shown in figure 5.1, with a process called an *axon*, which functions as an output for the unit, and a tree like structure, formed of *dendrites*, which functions as an input structure. Between the axon output, and the dendrites, there is the cell body, or *soma*, which also contains support structures needed in any living cell. Sometimes axons branch into several sections, as shown:

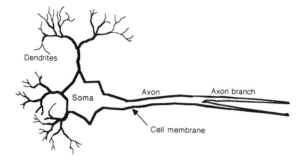

FIGURE 5.1   Schematic view of a neuron

Some neurons also have a specialised form of axon (figure 5.2), which is surrounded by a *myelin* sheathing, that has gaps at periodic intervals called *nodes of Ranvier*.

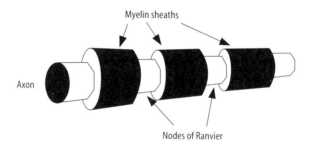

FIGURE 5.2   Myelinated axons

These axons are capable of transmitting information over relatively large distances with less expenditure of energy than unmyelinated axons. It is the break down of myelin which causes neurological problems in diseases such as multiple sclerosis in humans, because it prevents the long distance signalling between neurons. Functionally, neurons behave as processing units, but they are also able to amplify and transmit signals over large distances.

The connections between neurons are by means of *synapses*, which are normally considered to be between an output axon of the *pre-synaptic* cell and an input dendrite of the *post-synaptic* cell. These are called *axonal-dendritic* synapses. Other forms of synapse may also exist, such as *axonal-somatic* etc. In a chemical synapse the behaviour of the synaptic connection is determined by neuro-transmitters and the effect

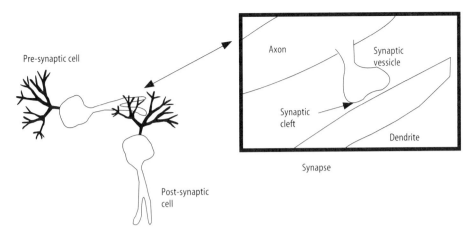

FIGURE 5.3 Synapses

of the synapse on the post-synaptic cell may be *excitatory* or *inhibitory*, depending on the neuro-transmitter within the synapse and its relation to the binding receptor sites on the post-synaptic cell. Figure 5.3 shows the structure of a typical synapse.

The firing behaviour of a typical neuron is characterised by two *passive* forms, *hyperpolarisation* and *depolarisation* (modest) and an *active* form in which the neuron *fires* and exhibits an *action potential*. The neuron is enclosed by a *membrane* which encompasses a naturally occurring electrical potential called the *resting potential*. The membrane potential can be modified artificially within the axon and if the potential is increased (hyperpolarisation), the response is a passive following of this increase along the axon. Similarly, if the potential is decreased slightly (depolarisation), the response is a passive following of this decrease along the axon. However, if the potential is decreased significantly, so that the potential inside the axon is above a *threshold potential*, then the neuron fires, with a characteristic action potential. Further if the stimulus potential is maintained, then a characteristic pattern known as a *spike train* is produced. The frequency of the spike train is usually proportion to the strength of the input stimulus. Some neurons adapt with continued stimulus, the frequency of the spike train will decrease with time and the neuron may eventually cease firing. These behaviours are shown in figure 5.4 below and discussed in greater detail in Stevens (1966), Dowling (1992) and Alberts *et al.* (1983).

McCullogh and Pitts described a simple formal model, based on the biological cell, which is equivalent to a *Threshold Logic Unit* (TLU). This can be described in terms of the equation:

$$y = H^* \left(\sum_{i=1}^{N} x_i w_i - \theta \right) \tag{1}$$

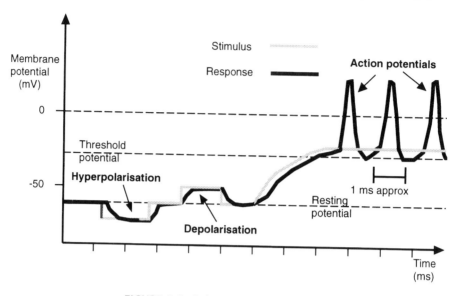

FIGURE 5.4 Behaviour of neuron when stimulated

where x_i are the inputs to the neuron, y is the output, and w_i are weights, q is a threshold, N is the number of synapses, and H^* is an extended Heaviside[1] function, defined by:

$$H^*(x) = \begin{cases} 1, & \text{if } x \geq 0 \\ 0, & \text{if } x < 0 \end{cases}$$

The weights are conventionally taken to be positive for excitatory synapses, and negative for inhibitory ones.

Equation 1 can be re-arranged to form:

$$y = H^*\left(\sum_{i=0}^{N} w_i x_i\right) \tag{2}$$

where conventionally x_0 is a bias input and set to 1 and w_0 is the associated weight, which is equal to $-\theta$.

The quantity:

$$\sum_{i=0}^{N} w_i x_i \tag{3}$$

is usually called the *activation* of the neuron and is a representation of the total stimulus applied.

Figure 5.5 shows the structure of a TLU. Note that this neuron model does not explicitly model time dependent behaviour and it is only an

---

[1]The Heaviside function is not normally defined for x = 0

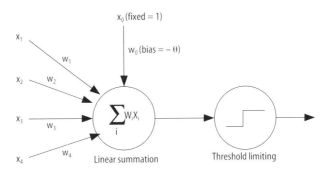

FIGURE 5.5 Schematic of a TLU

approximate model, which demonstrates the gross behaviour of real neurons. It does not model more complex behaviour, or any adaptation properties which real neurons possess. To many workers, who work only with artificial neurons, real neurons are simple, but biologists working with natural neurons know that there are many different types, each with different structure and behaviour. One common real behaviour is adaptation, which results in the strength of a neuron's output response diminishing over a period of time when the input stimulus is maintained.

Hodgkin and Huxley (1952) went on to develop a more accurate set of differential equations, based on their studies of giant squid axons – it was the axons which were giant, not the squid – since the experimental techniques available at the time would not permit investigation of neurons with small axons. There are still workers producing more detailed neuron models, which attempt to capture the detailed behaviour of biological neurons. Many of these models are based on electrical analogues – for details see Koch and Segev (1989).

Rosenblatt (1962), developed artificial models based on artificial neuron models which he called *perceptrons*. He built up networks of such units and demonstrated that they could perform useful operations. However, there was no mathematical proof that such networks would behave in predictable ways, and Minsky and Papert attempted to set these networks on a more formal basis, which led to the publication of their critical work on perceptrons in 1969. This was generally interpreted in a negative way and is held to have retarded the development of neural networks as a subject for study.

Some workers, such as Widrow, continued work on models of learning systems. Widrow's Adaline model was in fact a simple neural model, and the LMS algorithm of Widrow (Widrow and Hoff, 1960) forms the basis of modern training algorithms, such as *back propagation*.

Although many workers lost interest in neural systems following Minsky and Papert's negative results, some carried on. Little and Shaw produced models of feedback networks (Little, 1974, Little and Shaw, 1974), which were relatively unknown until John Hopfield published his reports of a less comprehensive system in 1982. Kohonen worked on models of associative memory (Kohonen 1977, 1984), and developed both unsupervised and supervised raining algorithms, which exhibit some biological plausibility. Grossberg has been working on biological models since 1970 and this has led to the development of network systems which exhibit attentional mechanisms (ART1, ART2 ARTMAP etc).(Carpenter, Grossberg and Reynolds, 1991 and Carpenter, Grossberg and Rosen, 1991).

Significant breakthroughs came in the 1980s, when Rumelhart, Hinton and Williams (Rumelhart et al, 1986) popularised the back propagation algorithm for training feed-forward networks. Subsequently it was shown that this algorithm had been developed by Werbos in 1974, and independently by Parker (Parker, 1985).

Other workers developed hardware related networks, with Stuart Kauffmann, 1969 working on Boolean genetic networks and Igor Aleksander developing hardware prototypes based on random access memory units, following the ideas of n-tuple recognition due to Bledsoe and Browning (1969). These models appeared to have less biological plausibility than models based on TLUs, but could be made to perform well in some pattern recognition tasks.

The current state of neural network research is that there are now a number of recognised artificial network architectures each capable of useful application. Applications which artificial neural networks are called upon to perform include:

- classification;
- image processing;
- auto association;
- hetero association;
- control functions.

5.3 FEED-FORWARD NETWORKS

The simplest form of artificial network is just a single artificial neuron. If the neurons are simple threshold logic units, they can be trained *by example* using a *supervised* training process. In this method, suppose that there is a set of patterns $\{P_i\}$, to associate with two output classes represented by 0 and 1. The patterns P_i are usually represented by a vector, which could contain integer, real, or complex components. In order to classify the patterns, a training procedure is adopted. First, the patterns are put into a suitable form for presenting to the neuron. For a single neuron, it would be normal to represent the pattern vectors P_i as vectors of binary values – say P_i', which is usually achieved by a simple mapping process.

Thus there is a coding operation which takes P_i to P'_i. For example if:

$$P3 = (2.75, 4.2, -3.6)$$

then the component 4.2 could first of all be represented by the number 42 (a simple multiplication by 10) and then by the binary pattern 0101010. This coding is a precise coding of the real number 4.2, but applying the same coding to 2.75 it may be necessary to lose information – thus

$$2.75 \rightarrow 27.5 \rightarrow 27 (\text{by truncation}) \rightarrow 0011011$$

Finally, negative numbers can be represented also, thus:

$$-3.6 \rightarrow -36 \rightarrow 1100100 \text{ (where the leading 1 is a sign bit)}$$

Note that this is not the only form of coding which can be used.

Sometimes *thermometer coding* is used – for example, consider the 32 bit coding for 1.8:

$$< \text{———————— } 32 \text{ ————————} >$$
$$1.8 \rightarrow 18 \rightarrow 00000000000000011111111111111111$$
$$< \text{——— } 18 \text{ ———} >$$

It will be noted that thermometer coding uses very long codes, although for some applications this is helpful.

Sometimes real numbers are simply coded up as themselves (using mantissa and exponent representation in binary) and complex numbers can be coded up as real number pairs.

Lastly, the operation of coding a pattern P_i to P'_i may also include pre-processing operations, which are specific to the application domain. Thus for image recognition applications, a possible pre-processing operation might be to convert pattern P_i into its Fourier transform before coding it for presentation to the neural network. (see Chapter 7 for details of Fourier transform.)

Although relatively little work has been done in this area, it would seem that the coding of data before presenting it to a neural network may have a significant effect on the quality of the results obtained.

Reverting back to our discussion about training a single neuron – let us suppose that the input pattern P_i has been converted into a vector P'_i which contains binary values 0 and 1 only. Our intention is to train the network to dichotomise patterns into the two classes represented by 0 and 1. It may be possible to carry this out by the following procedure:

1. Choose a pattern P'_i for training;
2. Apply P'_i to the neuron, and note whether the output is 0 or 1;
3. If P'_i is of class 1, and the output is 1, then there is nothing to do;
   If P'_i is of class 0, and the output is 0, then there is nothing to do;

If $\mathbf{P}_i'$ is of class 1, and the output is 0, then the total activation of the unit needs to be increased, so that the unit will have an increased tendency to fire. To do this, it is only necessary to modify the weights of the neuron which are connected to non-zero inputs.

A simple strategy is just to add a constant value to each of these weights, say 1 for example. A little thought reveals that since there are as many neuron weights as inputs, that the weights of the neuron can be modified by the rule:

$$w_j = w_j + P_{ij}'$$

where now P_{ij}' denotes the j th component of pattern $\mathbf{P}_i'$;

Similarly, if $\mathbf{P}_i'$ is of class 0 and the output is 1, then the total activation of the unit needs to be decreased so that the unit will have a reduced tendency to fire. The rule in this case is:

$$w_j = w_j - P_{ij}'$$

4. The training process continues by repeating steps 1–3, until all training patterns output the correct classification code 0 or 1.

This simple training procedure illustrates several properties which also apply to other, more complex, forms of artificial neural network:

A. The training proceeds by showing the examples to the network – which in this case is just a single neuron;
B. Error correction is applied when the network misclassifies;
C. The process is iterative and patterns may be shown to the network many times before training is complete;
D. Training is not guaranteed, since the algorithm given here may cycle indefinitely. In this case it is not possible to associate each representational patterns $\mathbf{P}_i'$ with its output classification code.

The weight update rules given in step 3 of the algorithm can be generalised to:

$$\mathbf{w} = \mathbf{w} + (t - y)\mathbf{P}_i'$$

where $\mathbf{w}$ is the weight vector of the neuron, t is the desired or target output of the neuron and y is the actual output. Note that the value of $t - y$ is in the set $\{-1, 0, 1\}$ since the neuron used includes the Heaviside limiting function.

With this vector notation, $\mathbf{w}$ represents the complete set of weights, and $\mathbf{P}_i'$ represents the complete coded pattern.

This algorithm is known as the *perceptron training rule*, and has the somewhat dubious distinction of working well, when it works, or not working at all! This property was discussed in Minsky and Papert's book, and is a property not of single patterns, but of the whole pattern set. When a pattern set can be learnt by this algorithm, the patterns are said to be *linearly separable*. In general, pattern sets are not linearly separable and this algorithm will not terminate. However, some pattern sets are linearly separable and the algorithm works. Thus, it may be possible to apply the following procedure for learning a pattern set:

> Apply the perceptron training algorithm with a limit on the number of pattern presentations. If the algorithm succeeds, then the patterns can be classified by the trained neuron.

This procedure will often give useful results, though may also result in failure.

It was this failure situation which presented a significant problem during the 1970s. It was thought (correctly, as we now know), that using a network of such neurons, rather than a single neuron, would overcome the problem. The network structure adopted to overcome this difficulty is the feed-forward network, shown below in figure 5.6.

Here the inputs are connected to neurons which feed their outputs forward through the network to an output unit. It is convenient (though slightly less general) to restrict the topology of the network so that intermediate units belong to layers of units. Such layered feed-forward networks are called *multi-layer perceptrons* or **MLP**s. The problem with these network structures is that it was not obvious how to train them.

Suppose a pattern $\mathbf{P}'_i$ is placed on the network and its output coding is incorrect. Which weights in the network should be adapted in order to correct the network's behaviour? This is known as the *credit assignment* problem – and since it refers to a location within the topological structure, we call it a *structural* credit assignment problem.

The solution, which Werbos (Werbos, 1974) discovered and which was later rediscovered and popularised, is to modify the behaviour of the

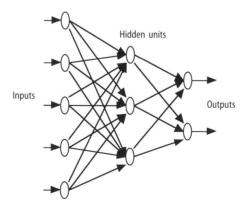

FIGURE 5.6   A fully connected feed-forward network (5:3:2)

individual neuron model. The simple formal neuron model has an input output function which is neither continuous nor differentiable when the total activation is zero – because of the Heaviside function. Suppose we modify the neuron so that the Heaviside function is replaced by another function which has a similar behaviour, but which is continuous and differentiable everywhere. Such a function is a sigmoidal function, such as:

$$\sigma(x) = \frac{1}{1 + e^{-kx}}$$

where k is a constant.

Notice that when k is very large, this function is a good approximator to the Heaviside function, but that it remains continuous and differentiable everywhere. In many artificial neurons, k is conventionally given a value of 1. By using such sigmoidal units in artificial networks, it turns out to be possible to overcome the credit assignment problems and this led to the adoption of the back propagation algorithm.

The back propagation algorithm is one of the most popular supervised training algorithms for training artificial neural networks. Essentially it works by defining an error function for the application problem and attempting to minimise the error by using a gradient descent optimisation algorithm. Because of the nature of the gradient descent optimisation used, it frequently gives slow training, and requires careful selection of a preselected parameter, called the *learning rate*. Choosing a learning rate for the back propagation algorithm is often problematic and may be considered to be a black art. If the learning rate is too small, convergence will be very slow, while if the learning rate is too large, the training process will oscillate. Often the learning rate is determined by trial and error, and careful monitoring of the error during the training process.

Besides its slowness, a major disadvantage of the method is that it is not guaranteed to find *global* solutions to the minimisation problem, but will normally find a *local* solution.

Other methods of training for feed-forward networks which may be considered are less frequently used, although this is probably because of lack of general awareness, rather than because the methods are not appropriate. These methods include:

- Random optimisation, which may give rapid convergence to global solutions (see Solis and Wets (1980); Baba (1989))
- Genetic algorithms, which may give rapid convergence to global solutions (see Goldberg (1989))
- Simulated annealing, which gives convergence to global solutions (see Geman and Geman (1984))

The random optimisation method in particular is easy to implement and could be expected to out perform back propagation, both with respect to the number of training cycles and also to obtain global rather than local solutions. However, there is no guarantee that global solutions will be found quickly in any method and Baba's method, based on the use of Gaussian distributions, which is known to be a global search method, may possibly be slower than Solis and Wet's suggested method of running a sequence of rapid local searches and selecting the best of them to find the global solution.

5.4 APPLYING NEURAL NETWORK METHODS

Many of the neural network training algorithms work in the following general way:

1. Choose a network topology;
2. Randomly, or otherwise, initialise the weights.;
3. Select a data set of representative patterns (the *training set*);
4. Determine an error function to be applied to the network and its data – this is frequently the sum of the squared error on the outputs over all the patterns in the training set;
5. Present patterns in some sequence (either in order, or randomly, or in proportion to their frequency of occurrence) and use the presentation of patterns to determine an error value for the network;
6. Use the error values to determine a procedure for varying the network parameters (weights) so as to reduce the error;
7. Repeat the steps 5 and 6 until either the error is sufficiently small, or some other terminating condition applies.

One problem with this method is that it reduces the error on the training set, yet what is often required is the minimisation of error on *unseen* patterns. A network which gives a good response on unseen patterns is said to *generalise* well. The process can give rise to *overtraining* if the error on the training set is reduced at the expense of generalisation. One solution to this problem is to use a second set of patterns as a *validation* set and to continue to train until the error on the validation set starts to increase. However, sometimes there is a restricted number of patterns available in the data set and in this case methods such as *cross-validation* can be used, which enable all the data to be used in training, yet still obtain a valid training for the network which will not overgeneralise.

Another problem which arises in many training algorithms is the number of presentations of each pattern which have to be made in order to train the networks. This is frequently very large, which results in a significant computational effort for training, and mitigates against training in real time. It is also not very biologically plausible, for those who worry about such things.

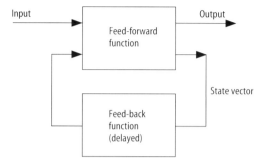

FIGURE 5.7   Model of feed-back system

Some algorithms, such as Grossberg's ART models (Carpenter, Grossberg and Reynolds, 1991 and Carpenter, Grossberg and Rosen, 1991) use attentional mechanisms in order to improve the training performance and to avoid or reduce the need for multiple pattern presentations.

5.5 FEED-BACK NETWORKS

Feed-back networks are more general than feed-forward networks and may exhibit different kinds of behaviour. A feed-forward network will normally settle into a state which is dependent on its input state, but a feed-back network may proceed through a sequence of states, even though there is no change to the external inputs to the network.

Feed-back networks can thus be used to model generators for oscillators, or to model time dependent behaviour.

A convenient model for a feed-back network is well known to engineers and has the form of the block diagram shown in figure 5.7. In this diagram the system has a set of external inputs, together with a set of outputs. There is also an associated *state vector* which describes the current state of the system. The feed-back connections are subject to a time delay, and this gives rise to time dependent output behaviour.

5.5.1 Lateral Inhibition

One of the simplest forms of feed-back network is the lateral inhibition network, which has been found to exist in biological systems. These networks were studied extensively by Rattliff and Hartline, in their study of the eye of the Limulus (Horseshoe Crab).

The network can be represented diagramatically by figure 5.8.

Figure 5.8 only shows lateral connections between nearest neighbours – in practice the inhibiting connections extend over a local neighbourhood.

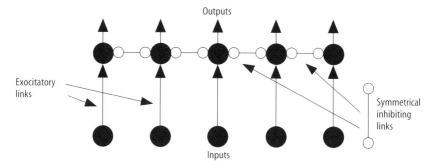

FIGURE 5.8   A recurrent laterally inhibiting network

Because of the lateral connections this is a feed-back network and it therefore has a time dependent behaviour. It acts as a *temporo-spatial filter* and will behave as a high pass spatial filter at low temporal frequencies, but will tend to become a low pass spatial filter as the temporal frequency increases. Such networks are often used in animal eyes, most probably as edge detectors or for contrast enhancement (where the temporal frequency is assumed to be low), but they could also be used as 'blob' detectors for moving objects, which would correspond to higher spatial frequencies. A carefully tuned lateral inhibiting network could function as a detector for objects moving across the visual field at a well defined rate and such networks may be used for this purpose in animals such as frogs.

The effects of a lateral inhibiting network for a constant image (zero temporal frequency) are shown in figure 5.9, which shows the contrast enhancement phenomenon.

The same effect can be achieve by a feed-forward lateral inhibiting network as shown in figure Y, but in this case the network functions only as a spatial filter, not as a temporal filter.

5.5.2 Hopfield Networks

Hopfield (1982) described a particular form of feed-back network. The characteristics of such a network are:

- Units do not self excite or inhibit themselves;
- The weights between the units are symmetrical;
- The units update asynchronously;
- The units are threshold logic units.

He showed that networks constructed in this way would always drop into stable states, irrespective of starting state and the order in which the nodes are evaluated. However, it is possible that the same network would drop into different stable states from the same starting state, if the evaluation order of the nodes is changed.

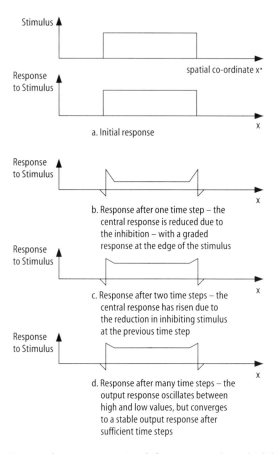

FIGURE 5.9   Temporal response to stimuli for recurrent lateral inhibiting network

He also associated a *Liapunov function* or '*energy*' with the states of the network and showed that this must always either remain the same, or decrease as each node is evaluated. Since the energy is bounded below, this implies that the network must become stable.

Hopfield suggested a *storage prescription* for patterns, so that the energy for the network when in a state corresponding to such a pattern would be low. The intention was that if the network were then started in a state sufficiently close to a stored pattern, the stored pattern would be retrieved. This notion allows us to use Hopfield networks as associative memories, which are conventionally of two types:

- Auto-associators;
- Hetero-associators.

Auto-associators essentially store patterns and give the input pattern as a response. In the limit these are just look up stores, which indicate whether or not a pattern has been stored (exactly) in the memory. Normally however, the retrieval process is not quite so exact and such auto-associators will respond with the stored pattern for input patterns which are close to the stored pattern. Such associators can therefore perform the following operations:

- Noise filtering;
- Pattern completion.

Thus, given a noisy version of a stored pattern, the associator will produce an idealised version. For pattern completion, if the associator is presented with a partial pattern, it will output a pattern in the store which matches the partial pattern adequately.

Hetero associators on the other hand, work with sets of pattern pairs – say $\{(P_1, Q_1), (P_2, Q_2) \ldots\}$. If the set of pairs is stored in the associator, then if a pattern P_i is presented to the associator, the result should be pattern Q_i. Some associators are also bi-directional, (for example Kosko's (Kosko, 1988) Bi-directional Associative Memory 1992), so that if a pattern Q_i is presented, then the output pattern will be P_i

Some associative memories function in a very general sort of way, so that if they are presented with either pattern (P_i^* or Q_i^* which are approximations of pattern P_i or Q_i they will output the pair of patterns (P_1, Q_1).

Within limits, auto-associators can be made to function as hetero-associators and vice-versa. A hetero associator can trivially be made to function as an auto associator, by simply storing the pairs $\{(P_1, P_1), (P_2, P_2) \ldots\}$. How well it will behave in its intended role will depend on how exact the pattern matching is in the particular model used.

An auto-associator, if large enough, can be used to simulate a hetero-associator. In this case, simply combine the patterns P_i, Q_i to be stored into a composite pattern $P_i + Q_i$, and then store the composite pattern. Providing the pattern completion properties of the associator are strong enough, then providing P_i (or Q_i) as partial input to the associator, will cause the whole composite pattern $P_i + Q_i$ to be generated as a response, from which the desired components can be extracted.

One problem with the use of Hopfield networks as associators is that they have a limited memory capacity. If the ratio of the total number of patterns stored to the total memory size is calculated, this tends asymptotically to zero as the memory size increases. However, for networks with around 1,000 nodes, the memory capacity has been shown to be approximately 15 per cent. As more patterns are stored, spurious stable states may arise in the network's state structure, which will give rise to unwanted associations and indeed this kind of memory typically suffers a rapid degradation in association performance once the memory limits are reached.

5.6 KOHONEN NETWORKS

Teuvo Kohonen worked on associative memories during the 1970s and also has been responsible for one of the most popular algorithms for artificial neural networks – the topological clustering algorithm, or *self organising map*. It seems possible that Kohonen was trying to develop his ideas regarding the adaptation of biological neural systems, by combining the behaviour of a network with laterally inhibiting connections with the application of an *unsupervised* learning rule – the so-called *Hebb rule* (Hebb, 1949).

Hebb's original statement of this rule has been interpreted in many ways subsequently, but essentially it states that if an input to a neuron is firing at the same time as the neuron itself is firing, then the connection (synaptic weight) between the neurons is increased. Note that this does not say anything about inhibitory synapses, or propose a means of decreasing weights, or propose any mechanism for the behaviour. Hebb's rule is often modified to allow for these behaviours.

This has led to the development of Kohonen's unsupervised training procedure, which is useful for pattern clustering. In this algorithm, all the units are linear and located on an organised hyper-surface, often a two dimensional grid.

The algorithm proceeds in the following way:

1. The network is initially configured with a randomly selected set of weights;
2. A pattern, P_i is chosen from the training set, and presented to the network. Normalisation may be required here;
3. The unit N_k which responds most strongly in the network is identified. This unit will have a weight vector which is close to the pattern vector for P_i;
4. A neighbourhood in the neural hypersurface is chosen around N_k. All the units within the neighbourhood are trained by a rule of the form:

$$\mathbf{w} = \alpha \mathbf{w} + (1 - \alpha)\mathbf{x}$$

   where $\mathbf{x}$ is the vector for pattern P_i, and $\mathbf{w}$ is the weight vector for each unit. α is a constant, in the interval $[0, 1]$ and usually close to 1 – say 0.9;
5. The algorithm is repeated from step two, although at each step the size of the neighbourhood may be reduced to aid with convergence.

When this algorithm is run, it is found that after training the units which respond to any pattern tend to form organised groups, with similar patterns clustered together and adjacent clusters having overlapping features.

Kohonen gives several examples, including the phoneme map (Kohonen, 1984; Kohonen, 1995). Kohonen has since gone on to modify his algorithms to form supervised algorithms, the so-called **LVQ** (*Learning Vector Quantization*) algorithms . These are an extension of the unsupervised algorithm, with the same network structure and generally work as follows:

1. First cluster the patterns, either by using k-means clustering, or by using the topological feature map method;
2. Now classify each neuron in the trained network. This can be done in the following way:
   a) associate a vector with integer components for each category in the training set and initialise all the components to zero
   b) present each pattern in the training set and locate the unit with the strongest response. Add one to the count of the appropriate category for that unit;
   c) the pattern class for each unit is now the category with the largest values in the class vector associated with the unit;
3. Train the units, by running through the training set one or more times and at each training step perform the following weight updates:

   a) **ATTRACT:**
      If the pattern is of the same class as the unit which responds to it, then move the weights closer to the pattern – usually done with an update rule of the form:

      $$\mathbf{w} = \alpha\mathbf{w} + (1 - \alpha)\mathbf{x}$$
      where α is in $[0,1]$

   b) **REPEL:**
      If the pattern belongs to a different class from the unit which responds maximally to it, then move the weights further away from the pattern – which may be done with an update rule of the form:

      $$\mathbf{w} = \beta\mathbf{w} + (1 - \beta)\,\mathbf{x}$$
      where $\beta > 1$

Variants of the LVQ algorithm use slightly different forms for the attraction and repulsion rules. For further details see Kohonen [Neural Networks etc].

The Kohonen algorithms are very useful. The unsupervised algorithms are useful for clustering the input data into groups, while the supervised LVQ algorithms are powerful enough to be used for classification without further processing.

When trying to solve classification problems, it is very useful to have a picture of the clusters formed by the data. Methods such as the backpropagation method for feed-forward networks are black box methods and provide little insight into the particular problem. Even if such networks are used to implement a solution to a classification problem, the insight gained by using the Kohonen algorithms on the input data can be considerable and can help to engineer a better solution.

Kohonen networks can also be used in hybrid solutions, where, for example, the input is first classified, and then an appropriate feed-forward network is selected for further training. There is some evidence that hybrid methods may give significant improvements for some applications.

5.7 SOFTWARE SUPPORT FOR NEURAL NETWORKS

Many neural network systems have been built from scratch, with researchers coding their systems in conventional languages. This is often necessary because ready made packages do not provide all the features which are required for a particular application. Further, by this approach, researchers become more familiar with the operation of the algorithms and this often leads to a better result. On the other hand, it is not always necessary for researchers to understand the detailed working of a particular neural network algorithm and the use of appropriate packages would considerably reduce the effort required to produce a suitable system. Further, making use of packages has the advantage of reducing the chances of coding errors and is likely to provide other tools which individual workers may find hard to code. There are a number of freely available software packages for neural network simulation which are likely to require some modification in order to realise a particular application model, but nevertheless the effort may be worth while. Particular systems which are still in use include:

- The Rochester simulator – a simulator package from the University of Rochester which handles basic ANN paradigms;
- The Stuttgart simulator – a simulator package from the University of Stuttgart, which handles several ANN paradigms.

There are also commercially available neural network toolkits and simulator packages, including:

- The Matlab neural network toolkit – a relatively simple package which is perhaps best suited for prototype examples and demonstrations;
- NeuralWorks – a commercial simulator package;
- NeuralWare – a commercial simulator package.

Commercial simulator packages are in a state of continuous development and no attempt is made here to evaluate what is currently available. However, researchers or network modellers with an application to implement should consider carefully whether it would be more cost effective to use a ready made package than to code their application from scratch. In research establishments the decision is often constrained by budgetary constraints and the desire of the researchers to understand and have complete control over the modification of the code, but for some applications where the emphasis is on applying standard techniques to a particular problem in (say) pattern recognition, this may not be appropriate.

5.8 HARDWARE SUPPORT FOR NEURAL NETWORKS

At one time it looked as though hardware support would be necessary for neural network applications, but continuing improvements in computer hardware has largely reduced the need for this. Many applications have real time classification or processing, but the training can be done off line. Modern computers are often capable of carrying out both the training and processing phases to a satisfactory level of performance.

Where very high levels of performance are required, there are now available a number of special purpose units – often available as special purpose cards for inclusion in a conventional computer system. These will typically contain signal processing chips in order to perform the vector and matrix multiplication operations which form the heart of many network processing algorithms. Some units may also exploit parallelism.

Another approach to providing hardware support is to use special purpose hardware chips. These may either be wholly digital in their operation, or may exploit analogue electronics. Some digital devices (see Mars), use pseudo random number generators to generate probabilistic bit streams as a representation for the continuous inputs to a neuron model – this gives rise to very fast switching operations, and very fast hardware operation.

Analogue systems for neural network modelling have been addressed by Carver Mead (1989).

5.9 IMAGE PROCESSING APPLICATIONS

Artificial neural networks can be used in image processing applications. A common application in clinical analysis is the scanning of slides, for example, to detect cell abnormalities in tissue. This process is often performed manually and is frequently error prone. The same techniques can be used in other research areas – particle physics, for example, to locate particle tracks from bubble chambers etc.

Many of the techniques used are variants of other commonly used methods of pattern recognition. However, whereas other methods of pattern recognition may require modelling of the objects to be found within an image, neural network models often work by a training process. However, such models also need attentional devices, or invariant properties, as it is usually infeasible to train a network to recognise instances of a particular object class in all orientations, sizes and locations within an image. One method which is commonly used is to train a relatively small window for the recognition of objects to be classified, then to pass the window over the image data in order to locate the sought object, which can then be classified once located. In some engineering applications this process can be performed by image pre-processing operations, since it is possible to capture the image of objects in a restricted range of orientations with predetermined locations and appropriate magnification. In general, however, this is not possible and the scanning window method must be used.

Systems have been proposed which perform image transforms before the recognition stage – these transformations can include performing Fourier transforms, or using polar co-ordinates or other specialised coding schemes, such as chain codes. One interesting neural model which has been developed intensively is the Neocognitron model of Fukushima (1988), which is capable of recognising characters in arbitrary locations sizes and orientations, by the use of a multi-layered network.

Particular pre-processing operations which may be needed for image processing typically include setting the quantisation levels for the image (these may vary across the image due to uneven lighting), normalising the image size, rotating the image into a standard orientation, filtering out background detail, contrast enhancement and edge direction. Standard techniques are available for these and it may be more effective to use these before presenting the transformed data to a neural network.

5.10 STRATEGIES FOR DEVELOPING APPLICATIONS

The field of neural networks has become large and it would require some time to become expert in all the currently used network paradigms. Many applications are developed using feed-forward networks, simply because the workers involved know that it will probably work, have the tools available and, are unaware of suitable alternatives.

For problems which require pattern classification it frequently helps to develop both a Kohonen feature map, using the unsupervised algorithm, and then go on to use LVQ methods in order to classify the data input. Other clustering methods, such as k-means clustering can also be used for the same basic purpose – gaining insight into the data set. Frequently it happens that some data patterns are easy to classify and clustering the patterns will help to identify those for which classification will be easy and also those patterns which will require more training effort.

The Kohonen algorithms are usually very computationally efficient and it may be that LVQ methods will actually solve classification problems well enough.

However, if this does not produce a satisfactory result, the insight gained is likely to be very useful. Further, it may be possible to use a preliminary classification based on one of the Kohonen methods in order to produce a hybrid network, perhaps composed of several feed-forward networks, which will outperform any individual network. One approach would be to perform supervised training of each feed-forward network having already clustered the input into several classes.

It has already been suggested that the back-propagation algorithm is not the best algorithm to use for training feed-forward networks and application developers should consider the use of other methods – in particular, Solis and Wet's random optimisation methods seem very effective. If back-propagation is used, then consider the choice of learning rate carefully, or choose a method which dynamically adjusts the learning rate, such as that due to Jacobs (1988).

Further, although many neural network applications start with a randomised set of weights, there could be considerable benefit in deliberately coding up some of the neurons so that they follow known rules. Alternatively, rule based KBS systems could be used as pre-processors to a neural network – thus, the neural network would then try to capture additional behaviour not captured by a rule based KBS.

Selecting the network topology for a neural network is often problematic, though it is known that using more hidden units in a feed-forward network usually makes it easier to train, though at the expense of having more computation on each training cycle. This does, however, tend to lead to overtraining and some methods have been developed which work either by dynamically increasing the number of hidden units in the network being developed, or by using a large number of hidden units and performing pruning operations to remove unnecessary units from the trained system, Some experimentation is often needed to find an appropriate number of hidden units for such networks.

Once trained, it is desirable to find the effectiveness of each hidden unit. This can be determined by first considering units which have small weights to output units – unless they generate large output values (possible for linear units, though not for sigmoidal ones – they should have little effect on the output and can be removed. Also, some units tend to have either a very low output or a high output on average and this may indicate that they are not participating in a useful computation – though note that this is not an invariant rule – for example, an eight input AND function would have an average output of 0.0039, but might be a useful function for a problem involving eight or more input variables. Perhaps the only way to be sure that units are not contributing to the trained network, is to selectively remove units (or clamp their outputs to zero, if that is easier) and to monitor the (expected) increase in error which should result. Units which have little effect when removed could then be considered superfluous.

A more powerful method is to use regularisation techniques to constrain the network output and control the effective network complexity. A commonly used regularisation techniques is *weight decay* which penalizes large network parameters.

Whenever possible, trained networks should be analysed so that a greater understanding of their function is obtained. Another test that can be performed, is to see which hidden units respond to particular patterns in the training or validation sets. For an example of a systematic analysis of a trained network, see the work on detecting shape from shading by Lehky and Sejnowski (1988).

Training set size is also a concern and wherever possible it helps to have a large amount of data available for training, even if it is not all used. Cross validation methods may be necessary for training and testing where only small data sets are available.

It also helps to have some idea of the nature of the training patterns and their frequency of occurrence. If a few significant patterns occur rarely, some networks will tend to put a lot of effort into learning the correct behaviour for more frequently occurring patterns, which is not always desirable. Although this effect can be overcome by using an error function, which takes frequency of pattern occurrence for classes of patterns into account, this will still lead to slow training for infrequently occurring pattern types. One possible way of compensating this, is to bias the occurrence of such infrequently occurring patterns in the training set, so that they are over represented. This will normally be effective in shortening the training time considerably, though care must be taken to make sure that the network does not become trained on rogue data.

If possible, analysis and monitoring tools should be used during training, for example to check that the total network error is decreasing in the expected manner. However, note that some systems use a graphical display for representing such error curves and that these may slow down the training considerably. Once the expected form of training behaviour has been achieved, it may be better to turn all monitoring off!

In conclusion, it has to be said that neural network techniques do offer the potential for carrying out some image processing and pattern recognition tasks effectively, but that this potential has not yet been fully realised. Systems such as the WISARD system (Aleksander, Stonham and Wilkie, 1983) provide very rapid real time training, but a somewhat unreliable performance in classification, while systems based on feed-forward networks are not generally suitable for real time training. It seems most likely that progress will be made through the adoption of hybrid systems and the use of attentional mechanisms similar to those used in natural systems.

Part 2
Image Processing: An Overview

A. BARRETT

6 Introduction to Image Processing

A. BARRETT

The majority of definitive texts concerned with Image Processing require a whole book of around several hundred pages with accompanying illustrations to present the subject in its entirety.

Since we are combining two specialised areas in a single volume, as explained in the preface, space is a constraint on the quantity of material that can be presented. In the chapters on Image Processing as with those on Knowledge-Based Systems, specific areas have been selected to cover a range of activities that will provide the reader with details on some of the methods, and their application, currently in use.

6.1 BACKGROUND

Early examples of image processing were the transmission of digitised newspaper pictures sent by submarine cable in the 1920s. These however, were of a rather poor quality in comparison with today's standards, due to the limited resolution achievable at that time. Steady improvements in quality were made during subsequent years, although it was not until the mid 1960s when a combination of developments in computing, combined with a demand for improvements to images obtained as part of the American space program, that significant advances were realised. As the developments in computer technology increased, thus enabling the capability for storing and manipulating complex images, the areas of application have correspondingly widened as indicated by the list in section 6.2 below.

6.2 WHAT EXACTLY IS IMAGE PROCESSING?

For our purposes and for the purposes of this book the answer is:

Image processing may be regarded as the application of a particular process to an image in order to obtain another image.

The above definition naturally raises corresponding questions such as:

What is an image and why do we need to process an image in order to obtain another?

Fundamentally, we may consider an image to be a pictorial representation of a particular domain.
Examples are:

- Medicine (x-rays);
- Biology (Electron Micrographs of biological material);
- Satellite Date (Pictures of the Earth and its Environment);
- Commercial Documents (Legal and Insurance records);
- Archaeological Data (Pictures of Relics);
- Forensic and Video Data (Fingerprints, Identification Shots);
- Industrial Processing (Quality and Assembly Control Data);
- Any camera negative.

The purpose in processing images is because we need to enhance the pictorial information for subsequent human interpretation.

6.3 IMAGE REPRESENTATION

As an example of image representation, consider a simple photographic negative obtained by an ordinary camera (figure 6.1). On closer inspection, the negative or image shows that it is made up of variations in light and dark at different parts of the negative according to the density of the photographic emulsion at any chosen point. We can in fact quantify this variation in intensity across the negative to give us what is commonly known as the gray level distribution of the image. Since the image is planar, we represent this distribution by a function $f(x,y)$ where $f(x,y)$ is simply the intensity at a selected point having co-ordinates x,y. To establish the co-ordinate structure we adopt the convention that, the top left hand corner of the image corresponds to the origin and the x and y axes are the corresponding vertical and horizontal directions drawn from the origin (figure 6.1). Each point x,y is called a 'pixel' which stands for 'picture element'.

6.4 IMAGE ACQUISITION

As shown in figure 6.1 above, images in some cases can be captured or acquired using an everyday camera although for most applications other more complex devices are necessary to obtain the image in sufficient detail for analysis.

FIGURE 6.1 An Image: In this case a photograph showing the axis convention used in image representation

One such device is the microdensitometer, which is frequently used in the electron microscopy field.

An electron micrograph consists of a film transparency obtained by recording the distribution of electrons scattered by a specimen in the microscope. The transparency is subsequently placed in the micro-densitometer and the intensity distribution is measured by focusing a beam of light on the transparency at different points and recording, using a photodetector, the amount of light either transmitted or reflected at each point.

Another common device used in image acquisition is a CCD (Charge Coupled Device) camera or sensor. This type of sensor can be configured either as a line scan sensor or as an area sensor. A line scan sensor consists of a row of silicon imaging elements that have a voltage output proportional to the intensity of the incidental light. This output can then be converted into digital form for subsequent input into the computer. The scanner operates by scanning the two-dimensional image on a successive line basis starting at the top of the image and working down. An area sensor on the other hand, consists of a matrix of imaging elements and is capable of acquiring an image in a way similar to a video device.

Scanners my be hand-held or fixed and the choice of scanner is dependent upon the field of application and the accuracy required for the image in question.

The resolution or extent to which details of the image can be recorded vary with the choice of scanner and can typically extend from 100 to

1,000 dots per inch (dpi). For images of say 1000*1000 pixels the time taken to acquire the scan may be a little as three seconds.

In order to illustrate a simple example of a digital image, consider a case where a digitising device (scanner etc) is required to capture an image of 8*8 pixels which represent the letter 'T'. Given that our device has a output range between 0 and 7 and that the lowest intensity value is assigned the value 0, we expect the digitised image to look like Fig 6.2:

0	0	0	0	0	0	0	0
0	0	0	0	0	0	0	0
0	7	7	7	7	7	7	0
0	7	7	7	7	7	7	0
0	0	0	7	7	0	0	0
0	0	0	7	7	0	0	0
0	0	0	7	7	0	0	0
0	0	0	0	0	0	0	0

FIGURE 6.2

Although the above example is a simple one, it nevertheless illustrates how numbers can be assigned in the digitising process to represent specific shapes or characters.

6.5 IMAGE STORAGE

The mode or method of storage is largely dependent upon two factors – the size of the image and the extent to which the image needs to be available for processing and analysis. For example, an 8-bit image of 256*256 pixels requires around 60,000 bytes of storage to be available. Larger images such as those generated by satellite data may require several million ie several gigabytes for just a single image. However, this latter demand does not constitute a problem unless we are required to access the image on a frequent basis. In these circumstances the storage would have to be effective in meeting such a requirement.

The storage device which offers rapid retrieval is memory, although when accessing large images this mode of storage is not appropriate and other methods must be sought. Another device is the frame buffer, which is a board allowing access at video rates ie around 30 images per second.

More recently optical disk technology offers storage facilities for large images with capacities of around a billion bytes (gigabytes) on a single platter. Several of these platters can be combined in a single unit known as a jukebox, which can be made available locally or via networks and clearly provide a very powerful medium for storage and access (Roth (1990)).

When the processing and analyses have been completed the image is stored in archive mode either using magnetic tapes or optical disks. Optical disks are increasing in popularity particularly as the shelf life of magnetic tape is only around seven years.

6.6 THE HUMAN EYE AND RECOGNITION

The eye, due to its flexibility, is generally regarded as significantly superior to any camera system developed to date. The range of intensities to which the eye can adapt is of the order 10^{10} ie from the lowest visible light to the highest bearable glare. The basic components of the eye consist of a lens, a retina and an iris. The central part of the retina called the fovea, contains between six and seven million cones which are sensitive to colour and are connected directly with the brain via individual nerves. An image that is projected on to the retina is converted into an electrical impulse by the cones and then transmitted by the nerves into the brain.

Another crucial part in the structure of the eye are the rods. These are distributed across the surface of the retina and number between 75 and 150 million. The rods, unlike the cones, share nerve endings, and are not involved in colour vision, but are sensitive to light and dark. In fact the cones do not function in dim light.

The eye is flexible in that, unlike a fixed camera, it is able to adapt to different situations of luminosity by changing its structure. For example, the iris may contract or expand to control the amount of light that enters the eye. Despite the remarkable capabilities of the eye, the chapters that follow show that in the form in which we receive most image material, the eye is often unable to distinguish the relevant components from the image and that these components are only recognisable after the image has been processed.

6.7 USE OF COLOUR

In general, the eye is less able to discern gradual changes in brightness compared to gradual changes of colour. It is thought that the eye is only capable of differentiating between approximately sixteen distinct gray levels although Niblack (1986) suggests that up to fifty gray levels can be identified on most monochrome displays. With colour, however, the eye is capable of distinguishing around two hundred colour shades (Niblack (1986)).

Plate I illustrates the effectiveness of using colour in surgical imaging. The different colour components enables the eye to easily distinguish the different parts of the anatomy in the area under investigation. Similarly, Plate II shows the effectiveness of assigning different colours to differentiate the structural components in a cross section of a plant infesting organism.

The assignment of colour can be done in two ways. Either we can use a colour sensor such as a colour TV camera, or we can assign a particular colour to every gray level intensity value on a graphics display of the image. This latter approach assumes that the primary colours, red, green and blue (RGB) are each scaled to lie in the range 0–1 and that for a given gray level a proportion of each of the RGB components can be appropriately assigned. The method of assignment is based on what is known as the RGB model (Gonzalez & Woods (1992)) and assumes that the three RGB components comprise the axes of a unit cube. The diagonal of the cube represents the range of gray level intensities. The origin of the cube corresponds to the white end of the gray scale and has RGB component values (0, 0, 0). The opposite end of the diagonal represents black and has RGB value (1, 1, 1). Intermediate values on the gray level intensity diagonal can be assigned proportionate values of the RGB components.

6.8 IMAGE SAMPLING

Before digitising an image we need to know how many pixel gray levels we should sample so that a subsequent display of the levels shows the image in sufficient detail.

If we over sample then we have a situation in which the image contains redundant information implying that we would be taking unnecessary storage. On the other hand, if we under sample, a subsequent display of the image will show insufficient detail for us to arrive at a correct interpretation of the image. Figure 6.3 shows the effect of different sampling rates on an image. Figure 6.3 (a) shows a square image sampled on a grid of 512 by 512 points with 256 gray levels. Figure 6.3 (b) through to 6.3 (f) shows the same image but with the sampling grid successively halved.

The size of the display was kept at 512 by 512 points and to fill the screen, pixels in the lower resolution images were duplicated. The result clearly shows that below a certain level of sampling the image becomes uninterpretable.

Although there are no strict criteria for establishing an optimal sampling rate, as a general rule, a minimum system for most image processing work should be able to display 256 by 256 pixels with 64 gray levels (Gonzalez & Woods (1992). Also see next chapter on frequency space sampling.)

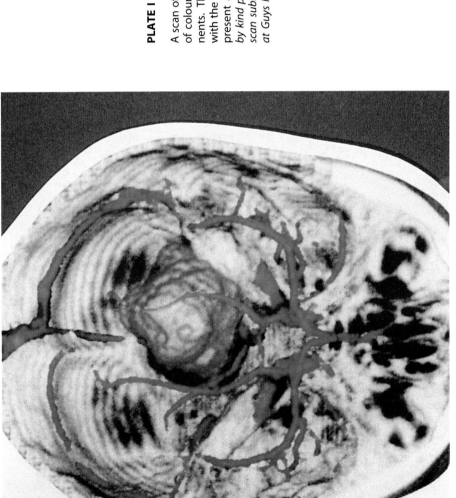

PLATE I

A scan of the brain showing the effectiveness of colour in representing anatomical components. The blood vessels are coded as red with the bone structure as white. A tumour is present and is shown in green. *(Reproduced by kind permission of Image Processing from a scan submitted by the Department of Surgery at Guys Hospital London.)*

PLATE II

Different colours are assigned to assist in the identification of the numerous structural components of the nematode, a plant infesting organism. *(Reproduced by kind permission of Dr. M. M. Jordan of the Department of Informatics National Institute for Biological Standards and Control.)*

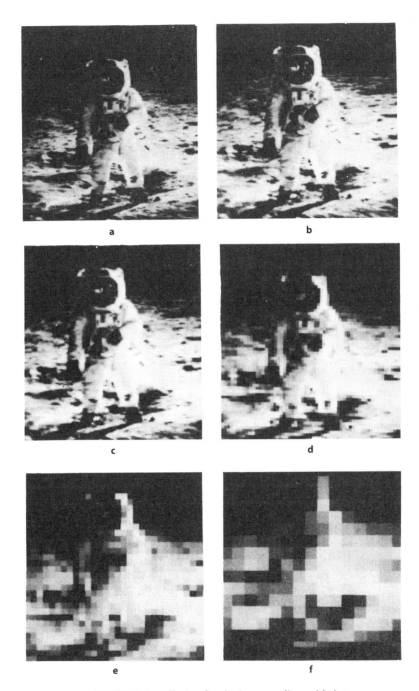

FIGURE 6.3 Effects of reducing sampling-grid size

6.9 IMAGE PROCESSING OVERVIEW

For a given image, we now define some of the basic steps which represent the overall procedures that combine to represent an image processing system:

1. Acquire the image in a digital form using a suitable digitising device;
2. Display all or part of the image depending upon the size of the image and the nature of the problem domain. (The display may be in either monochrome or colour);
3. Select an appropriate transformation process in order to enhance the image or to isolate regions specific to subsequent analysis;
4. Analyse and interpret the image and store the result together with the enhanced image.

The bulk of the processing is in the two latter steps and it is these steps with which the remainder of the book is concerned.

7 Frequency Space Analysis

A. BARRETT

7.1 THE FOURIER TRANSFORM

Before beginning our mathematical section on the transform, it is useful first to describe the situation in words.

If a problem is difficult, we try to 'look at in another way' or to 'turn it round so that we can see it differently'. The techniques of transforming a situation are basic to mathematics and there are very many different kinds of transformation. A picture, a distribution of density in a space of one, two or three dimensions may be transformed and represented in a new way in another space. One of the most important ways is a Fourier transformation (called after Jean Baptiste Joseph Fourier who first described the method in 1812 when dealing with problems in the flow of heat).

We may set out quantitatively the properties of the Fourier transform as follows:

(1) A density distribution can be built up by the linear superposition of a number of sine waves of different frequencies. If the density distribution is periodic, of frequency f (like the complex note of a violin), then these sine waves have only the frequencies f, 2f, 3f . . . nf . . . but if the density distribution is non-repeating, like a single pulse, then an infinite band of frequencies is required. Resolving a complex signal, into its Fourier components (the sine waves) is known as Fourier analysis. Combining the sine waves together again to give the general signal is correspondingly called Fourier synthesis;

(2) For each constituent sine wave we have to state its frequency, its amplitude and its phase (where the peak falls with respect to some reference signal). If the wave is in more than one dimension then we must also state its direction. As we will see, complex numbers provide an appropriate mathematical representation of amplitude and phase;

(3) In many physical systems, such as for radio waves in space or for light in a camera, the principle of linear superposition applies. This means that wave trains can cross each other without getting mixed

up. If linear superposition does not apply then two waves cross-modulate each other and new sum and different frequencies arise. Non-linear devices have their important place. If we have moving waves then it is vital to know whether waves of all frequencies travel with the same velocity (as is the case for electromagnetic waves in free space, but it is not so for a medium such as glass). We have to be clear on the properties of the particular system with which we are dealing;

(4) The spectrum of sine waves which make up a particular density distribution is know as its *Fourier transform*. In representing the transform we have to make provision for noting the amplitude, frequency, phase and direction of every constituent wave. We can plot out the transform in *transform* space. If the scales used in the *direct space* or *real space* in which we plot our original density distribution are, for example centimetres, then the scales used in the transform space will be reciprocal, for example per centimetre, which we may call reciprocal centimetres;

(5) When we buy an audio amplifier we ask about its frequency response and hope that all frequencies of notes in the complex sound signal going in, will come out 'undistorted'. The curve which describes the response of the system to different frequencies is called the 'contrast transfer function'. Asking about the contrast transfer function of an optical lens, an electron microscope, or of some other system has revolutionised our understanding of these instruments.

Many problems, then, are simplified if we first resolve a complex density distribution into sine waves and ask what happens to each sine wave separately as it passes through the system. We then re-assemble the somewhat changed sine waves to give a resultant density distribution. What happens to each sine wave may be represented in transform space and is the contrast transfer function mentioned above. Waves may be changed in amplitude, phase or direction. We may be considering the properties of a physical system, such as a microscope, or we may be using a computer as a kind of generalised imaging system, where we can apply any contrast transfer function, physically realisable or not;

(6) We have to build up a kind of library or dictionary of typical density distributions and their transforms. Such lists, in mathematical form, are to be found in handbooks. Visually, a picture in real space corresponds to another picture in transform space. It is important to acquire a qualitative pictorial appreciation of a range of such relationships to understand what is happening mathematically. The dictionary is reciprocal and works equally in both directions – transforming a transform takes us back to the original distribution;

(7) There is an important theorem or principle, the *convolution theorem*, which helps us to build up this library. It tells us how to think about the transforms of complex patterns in terms of the

transforms of simpler objects. The concept of *convolution* is required. If we have two density distributions A and B (two pictures for example), then the convolution of A and B means repeating the whole of A at every point of B, or vice versa (it does not matter which way we go). For example, if A is a circle and B is a lattice of points, then the convolution of A and B (which we may write as conv (A, B)), is a lattice of circles;

The convolution theorem then states that if we have a distribution in real space which is the point-by-point product of the distributions A and B, then the transform of this distribution is the convolution of transform (A) with transform (B) (see section 7.5). The theorem works in either direction. If in real space the distribution C is the convolution of the distributions A and B, the transform (C) is the point-by-point product of the transform of A and the transform of B;

(8) The convolution theorem enables us to do many remarkable things. During the Apollo 13 space flight the astronauts took a photograph of their damaged spacecraft, but it was out of focus. It is possible by image processing methods to put such an out-of-focus picture back into focus and thus to clarify it (see section 7.5).

Before formulating the mathematics of the Fourier transform an example is given to illustrate the power of Fourier processing methods applied to images containing a periodic substructure. In this example the application is to a biological image.

One of the most widely used techniques for the study of biological materials, such as viruses or components of isolated cells, is electron microscopy. This process involves exposing the biological material to a beam of electrons which are then scattered by the material. The subsequent pattern of the scattered electrons is recorded photograph-ically and it is this pattern which provides details about the structure of the material. The photograph is referred to as an electron micrograph and can be automatically converted into a set of digital density values by assigning numbers to the different levels of intensity within the micrograph. Figure 7.1a shows an electron micrograph of material comprising one of the surface proteins which forms part of the structure of the influenza virus. Figure 7.1b shows an enlarged display of the digitised micrograph. This digitised set of densities is stored in the computer, the Fourier transform applied and the corresponding frequency components displayed (figure 7.1c).

The figure, however, shows two types of distribution; one which consists of larger spots with a regular spacing between them, and the other which consists of smaller spots with no apparent regular spatial features. The former set of spots correspond to the basic frequencies of the material and the latter arise from the 'noise' components which occur during the process of obtaining the electron micrograph.

The 'noise' constitutes an unwanted part of the micrograph, since it obscures much of the structural detail and we must therefore try and remove it. We achieve this by retaining only those spots in the frequency display which correspond to the basic structural frequencies and then apply the inverse Fourier transform (see section 7.7) to arrive at a 'noise free' image. The most striking feature of this process, called digital spatial filtering, can be seen in the comparison between figures 7.1b and 7.1e, where the structural details of the image can be clearly seen, whereas they were previously obscured by 'noise'.

7.2 FORMULATION OF THE FOURIER TRANSFORM

The discrete one-dimensional Fourier transform of the function $f(x)$, evaluated over N points, may be written as :

$$F(u) = \frac{1}{N} \sum_{x=0}^{N-1} f(x) \exp(-2\pi iux/N)$$

for u = 0, 1, 2, . . . , N–1   with $i = \sqrt{-1}$

The transform has an inverse representation which may be correspondingly written as:

$$f(x) = \sum_{u=0}^{N-1} F(u) \exp(+2\pi iux/N)$$

for x = 0, 1, 2, . . . , N–1

In applying the transform in practical situations we frequently need to use both the transform and its inverse form, particularly in situations where modifications are made to the transform values and we require to see how these modifications may affect our original values $f(x)$.

For simplicity we have illustrated the one-dimensional case, although for the application of Fourier transforms to actual images we require the implementation of the two-dimensional form. This we can do with no loss of generality although we must remember that our two-dimensional function $f(x, y)$ now represents the distribution of gray levels as defined previously in Chapter 6.

The two-dimensional representation of the Fourier transform taken over an image whose x and y dimensions are each of value N is:

$$F(u, v) = \frac{1}{N^2} \sum_{x=0}^{N-1} \sum_{y=0}^{N-1} f(x, y) \exp\left[-\frac{2\pi i}{N}(ux + vy)\right]$$

for u = 0, 1, 2, . . . , N–1;   v = 0 , 1, 2, . . . , N–1

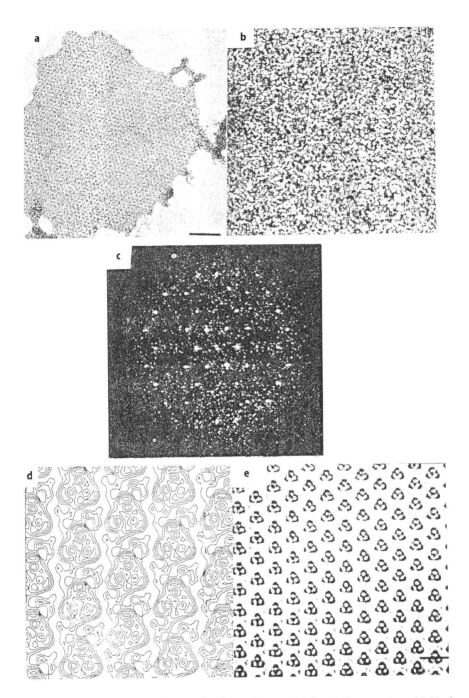

FIGURE 7.1 (a) Electron micrograph of protein material from influenza virus. (b) Display of digitised area within micrograph. (c) Computerised Fourier transform of (b). (d) Contoured display of selected area from 'noise free' image. (e) Display of 'noise free' image corresponding to area shown in (b)

The corresponding inverse relationship is:

$$f(x, y) = \sum_{u=0}^{N-1} \sum_{v=0}^{N-1} F(u, v) \exp\left[+\frac{2\pi i}{N}(ux + vy)\right]$$

for x = 0, 1, 2, . . . , N–1;   y = 0, 1, 2, . . . , N–1

In order to obtain a representation of the type shown in figure 7.1(c), we must evaluate what is called the Fourier spectrum.

Using the expansion for $\exp(-2\pi i \frac{ux}{N})$ as $\left(\cos 2\pi \frac{ux}{N} - i \sin 2\pi \frac{ux}{N}\right)$ the components of the above expression can be written as:

$$R(u) = \frac{1}{N}\sum_{x=0}^{N-1} f(x) \cos 2\pi ux/N$$

$$I(u) = \frac{1}{N}\sum_{x=0}^{N-1} f(x) \sin 2\pi ux/N$$

for u = 0, 1, 2, . . . , N–1

R(u) and I(u) are said to represent the real and imaginary parts of the transform.

The Fourier spectrum is then defined by the formula

$$|F(u)| = [R^2(u) + I^2(u)]^{\frac{1}{2}}$$

for u = 0, 1, 2, . . . , N–1

In two dimensions the corresponding formulation is :

$$|F(u, v)| = [R^2(u, v) + I^2(u, v)]^{\frac{1}{2}}$$

for u = 0, 1, 2, . . . , N–1; v = 0, 1, 2, . . . , N–1

Where:

$$R(u, v) = \frac{1}{N^2}\sum_{x=0}^{N-1} \sum_{y=0}^{N-1} f(x, y)\cos\frac{2\pi}{N}(ux + vy)$$

$$I(u, v) = \frac{1}{N^2}\sum_{x=0}^{N-1} \sum_{y=0}^{N-1} f(x, y)\sin\frac{2\pi}{N}(ux + vy)$$

For each value of u and v we determine our corresponding value |F(u, v)| which gives the spectral representation of the type shown in figure 7.1(c). In general images containing periodic or regular features whether visible to the eye or not, the spectral representation provides us with a highly effective means of representation of those features for subsequent analysis.

7.3 THE FAST FOURIER TRANSFORM (FFT)

Computation of the Fourier transform using a standard approach requires of order N^2 operations for N data points. This results from the fact that for every single value of u in:

$$f(u) = \frac{1}{N}\sum_{x=0}^{N-1} f(x)\exp(-2\pi iux/N)$$

for u = 0, 1, 2, . . . , N–1

we have to evaluate the summation over N data values, which gives a computational dependence of around N^2 in terms of the number of operations required. For many years, this made the routine application of the Fourier transform to large data sets a prohibitive undertaking. However, during the mid 1960s a method attributed to Cooley and Tukey (1967), showed that the computational dependence could be reduced from order N^2 to order $N\log_2 N$. Clearly for large N this yields a drastic reduction in terms of the number of operations and hence corresponding computer time. For example, for N = 256, N^2 = 65536, whereas $N\log_2 N$ = 2048. (Computationally a saving of around 63, 000 operations).

The method is based on the fact that the Fourier transform (Gonzalez & Woods (1992).may be written in the form:

$$F(u) = \frac{1}{2}\left[\frac{1}{M}\sum_{x=0}^{M=1} f(2x)W_M^{ux} + \frac{1}{M}\sum_{x=0}^{M=1} f(2x+1)W_M^{ux}W_{2M}^{u}\right] \qquad (7.1)$$

for u = 0, 1, 2, . . . , M–1

with $W_N = \exp(-2\pi i/N)$

N is of the form $N = 2^n$ where n is a positive integer. Based on this, N can be expressed as N = 2M where M is also a positive integer.

If we now further define:

$$F_{even}(u) = \frac{1}{M}\sum_{x=0}^{M-1} f(2x)W_M^{ux} \qquad (7.2)$$

for u = 0, 1, 2, . . . , M–1 and

$$F_{odd}(u) = \frac{1}{M}\sum_{x=0}^{M-1} f(2x+1)W_M^{ux} \qquad (7.3)$$

for u = 0, 1, 2, . . . , M–1

equation 7.1 may be written as:

$$F(u) = \frac{1}{2}\{F_{even}(u) + F_{odd}(u)W_{2M}^u\} \tag{7.4}$$

Also, since $W_M^{u+M} = W_M^u$ and $W_{2M}^{u+M} = -W_{2M}^u$ it follows from equations 7.1 to 7.4 that:

$$F(u + M) = \frac{1}{2}\{F_{even}(u) - F_{odd}(u)W_{2M}^u\} \tag{7.5}$$

Consider the above equations 7.1-7.5 in relation to a transform consisting of two data values ie $M = 1$. Then the above equations may be written:

$$F_{even}(0) = f(0)$$
$$F_{odd}(0) = f(1)$$
$$F(0) = \frac{1}{2}\{F_{even}(0) + F_{odd}(0)W_2^0\} = \frac{1}{2}\{f(0) + f(1)W_2^0\}$$
$$F(1) = \frac{1}{2}\{F_{even}(0) - F_{odd}(0)W_2^0\} = \frac{1}{2}\{f(0) - f(1)W_2^0\}$$

Computationally therefore, the two Fourier coefficients $F(0)$ and $F(1)$ are simply determined by firstly calculating $F_{even}(0)$ and $F_{odd}(0)$ which are just the data values $f(0)$ and $f(1)$ thus requiring no multiplications or additions. One multiplication of $F_{odd}(0)$ by W_2^0 and one addition yields the coefficient $F(0)$. Similarly $F(1)$ can be determined by a simple subtraction (computationally the equivalent operation to addition) of $F_{odd}(0)W_2^0$ from $F_{even}(0)$. Therefore the total number of operations required for a two point transform consists of one multiplication and two additions. It may be shown, using the principle of induction, that for $N = 2^n$ data values the number of complex multiplications and additions is $1/2Nn$ and Nn respectively. (Gonzalez & Woods (1992)).

The importance of the above illustrations is that it is theoretically possible by splitting our set of N data values into two point pairs, repeated application of equations 7.1-7.5 enables us to determine the total transform in a time proportional to $N \log_2 N$. Before application of the equations 7.1-7.5 we need to restructure our original set of data values into appropriate pairs using a method known as 'Successive Doubling'.

To illustrate the successive doubling approach consider the function $f(x)$ defined over the range of values for which $x = 0, 1, \ldots, 7$ to yield a set of values:

$$f(0), f(1), f(2), f(3), f(4), f(5), f(6), f(7)$$

The first step is to divide the array into so-called even and odd parts as follows:

even parts : f(0), f(2), f(4), f(6)

odd parts: f(1), f(3), f(5), f(7)

Each of the above represents an individual array which may be further subdivided into even and odd parts to give for the first array:

Even part : f(0), f(4)

Odd part : f(2), f(6)

And for the second array:

Even part : f(1), f(5)

Odd part : f(3), f(7)

Input into the above equations now consists of the set of two point transforms:

f(0), f(4); f(2), f(6);

f(1), f(5); f(3), f(7);

A constraint of the method is that the total number of data values must be expressible as a power of two. In practice this may not always be the case so in order to satisfy the constraint we simply 'pad' our data set to the next power of two with zeros. That is if our set corresponded to values for f(x) from x = 1, 2, . . . , 450 for example, we would simply generate values for f(x) all of zero magnitude for x = 451, 452, . . . , 512.

7.4 COMPUTING THE FOURIER TRANSFORM OF AN IMAGE

Since most images consist of a two-dimensional array of gray levels and we have only concerned ourselves so far with the one-dimensional determination of the Fourier Transform, we now address the question of extending our analysis into two-dimensions.

As shown in section 7.2, the form for the Fourier Transform in two-dimensions for a square or two-dimensional array is:

$$F(u, v) = \frac{1}{N^2} \sum_{x=0}^{N-1} \sum_{y=0}^{N-1} f(x, y) \exp[(-2\pi i / N)(ux + vy)] \qquad (7.6)$$

u = 0, 1, 2, . . . , N–1;   v= 0, 1, 2, . . . , N–1

However, since $\exp(-2\pi i/N)(ux + vy)$ may be written as $\exp(-2\pi i \frac{ux}{N})\exp(-2\pi i \frac{vy}{N})$ the above equation can be re-written as:

$$F(u, v) = \frac{1}{N^2} \sum_{x=0}^{N-1} \exp(-2\pi i ux / N) \sum_{y=0}^{N-1} f(x, y) \exp(-2\pi i vy / N)$$

Notice that the expression in the summation over y values is simply a one-dimensional transform for a fixed value of x over the frequency values v = 0, 1, . . . , N–1. This generates a set of Fourier coefficients F(x, v). The expression for F(u, v) may now be written as:

$$F(u, v) = \frac{1}{N^2} \sum_{x=0}^{N-1} F(x, v) \exp(-2\pi iux / N)$$

This in turn corresponds to a one-dimensional transform over the frequency values:

$$u = 0, 1, ..., N - 1$$

Therefore, we can regard our two-dimensional transform as consisting of a series of one-dimensional transforms which we would evaluate using the FFT algorithm.

Assuming the formerly defined axis convention for the image, (Chapter 6), the above representation means that the two-dimensional transform may be determined by firstly evaluating the coefficients F(x, v) from the transform of the image rows and then subsequently applying the transform of this (complex) set of coefficients along the image columns. (Note however that due to a translational property of the Fourier Transform, if we wish to display our spectrum of values with the origin at the centre so that the values are symmetrically disposed, we must multiply each gray scale value f(x, y) by $(-1)^{x+y}$ before invoking the transform process.)

As a further illustration of the use of the Transform and its relation to the set of spectral values defined in section 7.2, consider the image below (figure 7.2) which is taken from a central section of an image of 16 by 16 pixels.

The (*) symbol denotes a point in the image for which the gray scale value is greater than the background level (denoted as zero). The corresponding central part of the Fourier or Power spectrum obtained following the procedure outlined in section 7.2, would appear in the form shown in figure 7.2.

S denotes the sum of all the pixel or gray levels in the image and corresponds to the 1'st coefficient, often called the origin term, in the spectrum.

X denotes values within the spectrum (greater than zero) occurring for values of u and v and correspond in position to the reciprocal spacings between the periodic non-zero elements (*) in the original image.

Application of the inverse transform to the components used to construct the spectrum would obviously yield the original image.

IMAGE

```
*  0  0  0  *  0  0  0  * → y
0  0  0  0  0  0  0  0  0
0  0  0  0  0  0  0  0  0
0  0  0  0  0  0  0  0  0
*  0  0  0  *  0  0  0  *
0  0  0  0  0  0  0  0  0
0  0  0  0  0  0  0  0  0
0  0  0  0  0  0  0  0  0
*  0  0  0  *  0  0  0  *
↓
x
```

POWER SPECTRUM

```
0  0  0  0  X  0  0  0  0
0  0  0  0  0  0  0  0  0
0  0  0  0  0  0  0  0  0
0  0  0  0  0  0  0  0  0
X  0  0  0  S  0  0  0  X → v
0  0  0  0  0  0  0  0  0
0  0  0  0  0  0  0  0  0
0  0  0  0  0  0  0  0  0
0  0  0  0  X  0  0  0  0
      ↓
      u
```

FIGURE 7.2
An image subset and the
corresponding spectrum
obtained using the
Fourier transform

7.5 CONVOLUTION AND CORRELATION

Section 7.1 introduced the concept of convolution and referred to its importance in image processing applications. This section presents a theoretical explanation of the convolution process and the related process of correlation. The section also expands on the application of the Fourier transform in the evaluation of the above processes as introduced in section 7.1(7).

If we have two sequences $f(x)$ and $g(x)$, then the convolution of these sequences is expressed as $f(x)^*g(x)$ where:

$$f(x)^*g(x) = \sum_{x'=0}^{K-1} f(x')\, g(x - x') \tag{7.7}$$

where $g(x - x')$ corresponds to a translation x' of the function $g(x)$.

As pointed out in section 7.1 however, it can be shown mathematically that the above summation can be determined through the use of Fourier transforms, which is computationally more efficient than a straightforward evaluation, given the availability of the fast Fourier transform.

To achieve this requires the use of the convolution theorem, which states that the convolution of two functions, is simply the inverse transformation of the product of the Fourier transforms of the two functions. That is, if $f(x)$ has the Fourier transform $F(u)$ and $g(x)$ has the Fourier transform $G(u)$, then the convolution of the two functions may be written as:

$$T^{-1}\left\{F(u)\,G(u)\right\}$$

where T^{-1} denotes the inverse transform.

Convolution, as already mentioned, plays an important part in image processing applications. For example, if we have an image with gray-level distribution $f(x, y)$ and a function $h(x, y)$ which is designed to highlight some specific features of the image, we may write using the convolution theorem:

$$G(u, v) = H(u, v)F(u, v)$$

where $H(u, v)$ and $F(u, v)$ are the respective Fourier transforms of $h(x, y)$, and $f(x, y)$.

Then our new 'highlighted' image is simply:

$$g(x, y) = T^{-1}[G(u, v)]$$

Similarly, it is known that in certain images, degradation of the image takes place due to 'noise' or unwanted values which may result in blurring of the original image. If the 'noise' can be quantified and in certain image applications this happens to be the case, then we can say that our resultant image or the image that is actually recorded, can be represented as a convolution of the original image with a 'noise' function. Mathematically, if the Fourier transform of our 'noise' function is $N(u, v)$ say and $F(u, v)$ corresponds to the Fourier transform of our original image then:

$$g(x, y) = T^{-1}\left\{F(u, v)N(u, v)\right\}$$

What we have actually recorded is $g(x, y)$ and ideally we want to see the 'noise free' image $f(x, y)$. Therefore we may write:

$$F(u, v) = \frac{T\{g(x, y)\}}{N(u, v)} = \frac{G(u, v)}{N(u, v)}$$

Then the 'noise free' image ie $f(x, y)$, would result from the inverse transform of $F(u, v)$.

The above process which involves the division of Fourier transform coefficients is referred to as deconvolution.

To implement the above processes computationally requires some pre-processing of the data. This arises from the fact that in forming the Fourier products above for two arrays of different dimensions, errors in the higher order coefficients will result. This is due to the Fourier transform itself being a periodic function (Brigham, (1974)).

To overcome the problem requires that the value of K in the equation 7.7 is chosen to be consistent with the periodicity of the transform function. This requires (Brigham (1974)) that for two functions $f(x)$, $g(x)$ each of length N_1 and N_2 say, K should be chosen to be of a value such that:

$$K \geq N_1 + N_2 - 1$$

For practical purposes we generally choose K for which:

$$K = N_1 + N_2 - 1$$

This procedure therefore results in an extension to our original data dimensions and is achieved by simply appending zeros to each sequence as follows:

$$f(x) = \begin{cases} f(x) & 0 \leq x \leq N_1 - 1 \\ 0 & N_1 \leq x \leq K - 1 \end{cases}$$

and:

$$g(x) = \begin{cases} g(x) & 0 \leq x \leq N_2 - 1 \\ 0 & N_2 \leq x \leq K - 1 \end{cases}$$

Although the above analysis has been illustrated in relation to the one dimensional situation, the extension to two-dimensions means that instead of forming products between the Fourier coefficients for $F(u)$ and $G(u)$ we simply determine the products for $F(u, v)$ and $G(u, v)$.

7.6 EVALUATION OF THE INVERSE TRANSFORM

Modifications to the Fourier spectrum of any coefficients in the Fourier transform of the image, eg convolution of the image transform with a specified function, means that we will require to see the effect of such a modification on the original image.

This requires the use of the inverse Fourier transform to take us back from the domain of reciprocal space to image or real space.

This transformation as defined above may be written:

$$f(x) = \sum_{u=0}^{N-1} F(u) \exp(-2\pi iux / N)$$

$$x = 0, 1, \ldots, N-1$$

Taking the complex conjugate of the above and dividing both sides by N gives:

$$\frac{1}{N} f^*(x) = \frac{1}{N} \sum_{u=0}^{N-1} F^*(u) \exp(-2\pi iux / N)$$

(*denotes complex conjugate).

Compare this result with the expression for the Fourier transform ie:

$$F(u) = \frac{1}{N} \sum_{x=0}^{N-1} f(x) \exp(-2\pi iux / N)$$

The above comparison shows that the input of a set of values $F^*(u)$ into an algorithm designed to compute the forward transform yields the quantity $1/N \, f^*(x)$. Multiplication by N and taking the corresponding complex conjugate give the required values for $f(x)$.

7.7 SAMPLING

The periodicity of the Fourier transform imposes a constraint on our sampling interval of $f(x)$ if we are to successfully resolve the required information in the transformation process. This sampling criteria is known as the Whittaker-Shannon sampling theorem and states that if we require to resolve information within the function spaced at units distance 'd' apart, then we must sample our function at a minimum interval of 'd/2' units. In practice however, our sampling is generally chosen to be of a finer interval than the above, which is based on a 'noise-free' distribution of data values $f(x)$, as most data contains some noise due to imperfections which may result during recording of the original data.

8 Image Enhancement in Real Space

A. BARRETT

The previous chapter showed how we may modify and enhance an image by utilising operations in Fourier or frequency space. This chapter shows how modifications to the image may be made, for the purpose of enhancement through transformations applied to the individual gray levels of the image: a process sometimes referred to as 'gray level transformation'.

8.1 HISTOGRAM PROCESSING

A histogram of the frequency, or number of times, that a pixel with a particular gray-level occurs within an image provides us with a useful statistical representation of the image.

As an example, consider the image below which represents a dark object (a square on light background. The object is represented by gray levels greater than 4, and the background by levels between 0–2.

0	0	1	0	1	0	2	0
1	0	0	1	0	2	0	1
0	1	7	5	6	6	0	1
0	0	6	5	5	7	2	0
1	1	7	7	7	6	0	0
0	2	6	6	6	5	2	1
1	1	2	1	1	0	0	0
1	0	2	2	1	0	1	2

FIGURE 8.1
A square on a light background

The histogram representation of the above image is:

Gray Level	Gray Level Frequency
	0 5 10 15 20
0	* * * * * * * * * * *
1	* * * * * * * *
2	* * * *
3	
4	
5	* *
6	* * *
7	* * *

The above histogram is referred to as being 'bimodal' as it consists of two peaks: one in the region spanning gray levels 0–2 and the other spanning gray levels 5–7. Histograms only showing one significant peak are called unimodal.

In the case of complex images eg satellite or medical images which may consist of up to 256 gray levels and $3000*3000$ pixels, the resulting histograms will consist of many peaks and the distribution of those peaks together with the magnitude, can reveal significant information about the information content of the image.

Ideally, we would like an image whose histograms shows a uniform distribution over the range of gray levels covered, since this enables us to comprehend the contribution made by all the gray levels to the image. In practice, however, the resulting histogram is often non-uniform, with a few levels dominating over others to the extent that full image interpretation by the eye cannot be achieved. We therefore look for methods which transform the available intensities over a broader range. Figure 8.2(a) for example shows an image and its corresponding histogram is shown in figure 8.2(b). The horizontal axis in 8.2(b) represents a gray scale range from 0–255 and the vertical axis represents the frequency of occurrence of each gray level in the range. Note the 'bunching' of gray level values in the lower part of the range which corresponds to the rather dark appearance of the image. Figure 8.2(d) shows the gray level distribution after the application of a transformation, which effectively enhances the scale of values relative to the original scale. The validity of this approach to enhance the scale is based upon statistical theory and is explained in more detail in section 8.2.

Figure 8.2(c) shows the new image resulting from the gray level redistribution process. This process has increase the dynamic range of intensities to include more gray scale, with the result that the overall contrast is enhanced relative to the original image. This transformation process is called 'Histogram Equalisation' and the following section explains the method of implementation.

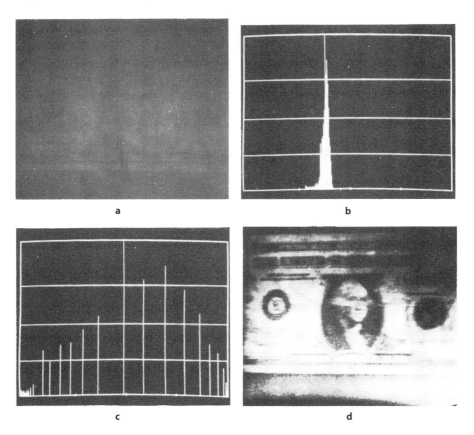

a

b

c

d

FIGURE 8.2 Illustration of the histogram equalisation approach (a) Original image (b) Original histogram (c) Equalised histogram (d) Enhanced image *(from Gonzalez and Wintz (1987))*

8.2 HISTOGRAM EQUALISATION

Ideally we would like our image histogram to be distributed across the range of gray scale values as a uniform distribution. This means that our contrast will be at an optimal level for subsequent image interpretation since all levels are represented. This is not generally the case in practice however and as shown above the distribution may be dominated by a few values spanning only a limited range. Statistical theory (Gonzalez and Wood (1992)) shows that by using a transformation function equal to the cumulative distribution of the gray level intensities in the image, enables us to generate another image with a gray level distribution having a uniform density.

The implementation of the above transformation can be summarised as a three stage process:

(1)   Obtain a histogram of the image;
(2)   Obtain the cumulative distribution of the gray levels;

(3) Replace the original gray level intensities by those determined in (2).

To illustrate the steps above we refer to the image defined in section 8.1 (figure 8.2) which represents a square on a light background.

The gray scale distribution of the individual levels is denoted by g_k where $k = 0, 1, \ldots, 7$ and is normalised by dividing the range by the maximum level, in this case 7. The frequency of occurrence of a particular gray level is denoted by f_k and S_k represents the cumulative sum which may be defined as $_R\sum f_R/n$ with n representing the total number of pixels in the image. The three quantities are represented in tabular form below:

TABLE 8.1 The distribution of gray level values for the image in figure 8.2. The corresponding normalised frequency values and the cumulative sum are also shown

g_k	f_k/n	S_k
0/7	22/64	22/64
1/7	17/64	39/64
2/7	9/64	48/64
3/7	0/64	48/64
4/7	0/64	48/64
5/7	4/64	52/64
6/7	7/64	59/64
7/7	5/64	64/64

Our S_k array now represents a new set of gray level values to replace the original values ie 0/7 above is to be replaced with 22/64. Similarly, the value 1/7 is replaced by the value 39/64 etc. However, since our range of allowable levels is from 0–7 (this may be the only range available to the display facilities), we need to equate our new levels to be compatible within this range. This done by simply rewriting both the g_k and S_k values in decimal form and then select those values from g_k closest to those in S_k:

TABLE 8.2 Decimalised values of the gray scale and cumulations

g_k	S_k
0.00	0.34
0.14	0.61
0.28	0.75
0.43	0.75
0.57	0.75
0.71	0.81
0.86	0.92
1.00	1.00

The closest value to S_0 ie 0.34 is g_2 ie 0.28 which in turn corresponds to the gray level value 2/7. Similarly, the closest value to S_1 is g_4 which corresponds to the gray level value of 4/7. Continuation of the process yields the corresponding gray level values for the set S_k as:

$S_k = (2/7, 4/7, 5/7, 5/7, 5/7, 6/7, 6/7, 7/7)$

for $k = 0, 1, \ldots, 7$

Substituting these values for those in the original image yields a new image as shown below:

2	2	4	2	4	2	5	2
4	2	2	4	2	5	2	4
2	4	7	6	6	6	2	4
2	2	6	6	6	7	5	2
4	4	7	7	7	6	2	2
2	5	6	6	6	6	5	4
4	4	5	4	4	2	2	2
4	2	5	5	4	2	4	5

FIGURE 8.3
The result of applying the equalisation process to figure 8.1

It should be emphasised that the above image is the result of applying the method to a frequency distribution in which the available levels were reasonably distributed and is only presented for the purposes of illustrating the method. The effectiveness of the approach is best appreciated by reference to figure 8.2 which represents an image of the type more likely to be found in practice.

8.3 HISTOGRAM SPECIFICATION

Although the above method enables the redistribution of gray levels over a uniform range, circumstances may arise where certain specific levels need to be highlighted relative to others. For example, even after application of the equalisation process, certain levels may still dominate the image to the extent that the contribution of the other levels cannot be interpreted by the eye.

One way to overcome this is to specify a histogram distribution which enhances selected gray levels relative to others and then reconstitute the original image in terms of the new distribution. Theoretical justification for this approach is explained fully in the book by Gonzalez & Woods (1992).

The implementation of the method can be illustrated in a similar way to that shown for the case of equalisation and consists of the three basic steps below:

(1) Apply the process of equalisation to the image;
(2) Specify a gray level distribution which enhances certain levels relative to others;
(3) Determine the compatible relationship between the levels determined in (1) with those determined in (2).

With reference to Figure 8.1 we repeat the results of the equalisation process showing the gray levels distributions in both fractional and decimal form:

g_k	f_k/n	s_k
0.7/ (0.00)	22/64	22/64 (0.34.)
1/7 (0.14)	17/64	39/64 (0.61)
2/7 (0.28)	9/64	48/64 (0.75)
3/7 (0.43)	0/64	48/64 (0.75)
4/7 (0.57)	0/64	48/64 (0.75)
5/7 (0.71)	4/64	52/64 (0.81)
6/7 (0.86)	7/64	59/64 (0.92)
7/7 (1.00)	5/64	64/64 (1.00)

In accordance with step 2 above we now specify a distribution with the appropriate properties. For example, we may decide to reduce the levels between 0 and 2 ie the background levels and increase the levels between 5 and 7 correspondingly. Suppose we further decide that the amount by which we choose to both correspondingly reduce and increase the frequency distribution is arbitrarily set as 7/64.

Then our new distributions may be written as:

g'_k	f'_k/n	S'_k
0/7 (0.00)	15/64	15/64 (0.23)
1/7 (0.14)	10/64	25/64 (0.39)
2/7 (0.28)	2/64	27/64 (0.42)
3/7 (0.43)	0/64	27/64 (0.42)
4/7 (0.57)	0/64	27/64 (0.42)
5/7 (0.71)	11/64	38/64 (0.59)
6/7 (0.86)	14/64	52/64 (0.81)
7/7 (1.00)	12/64	64/64 (1.00)

In accordance with step (3) above, we now equate the level S_k to those of S'_k as follows. The closest match to S_0 ie 0.34 is S'_1 ie 0.39 which corresponds to level 1/7. Similarly, the closest match to S'_1 ie 0.61 is S'_5 ie 0.59 which corresponds to level 5/7. Continuation of the process gives the following set of mappings

$$S_0 \rightarrow g'_1 = 1/7$$
$$S_1 \rightarrow g'_5 = 5/7$$
$$S_2 \rightarrow g'_6 = 6/7$$
$$S_3 \rightarrow g'_6 = 6/7$$
$$S_4 \rightarrow g'_6 = 6/7$$
$$S_5 \rightarrow g'_6 = 6/7$$
$$S_6 \rightarrow g'_7 = 7/7$$
$$S_7 \rightarrow g'_7 = 7/7$$

Insertion of these levels in place of those comprising the histogram equalised image give the new image shown below:

1	1	5	1	5	1	6	1
5	1	1	5	1	6	1	5
1	5	7	7	7	7	1	5
1	1	7	7	7	7	6	1
5	5	7	7	7	7	6	1
1	6	7	7	7	7	1	1
5	5	6	5	5	1	1	1
5	1	6	6	5	1	5	6

FIGURE 8.4.
The result of applying the specified gray level distribution to the image of figure 8.1.

To illustrate the effectiveness of the technique in a practical situation consider Figure 8.5(a) which shows a semi dark room viewed from a doorway. Figure 8.5(b) shows the image after histogram equalisation and Figure 8.5(c) is the result of interactive histogram specification. Figure 8.5(d) shows the histograms: from bottom to top, the original, equalised, specified and resulting histograms, respectively.

8.4 GRAY LEVEL THRESHOLDING

In section 8.1 we gave the histogram corresponding to Figure 8.1 as:

Gray Level (gk)	Gray Level Frequency (fk)
	0 5 10 15 20
0	* * * * * * * * * * *
1	* * * * * * * *
2	* * * *
3	
4	
5	* *
6	* * *
7	* * * *

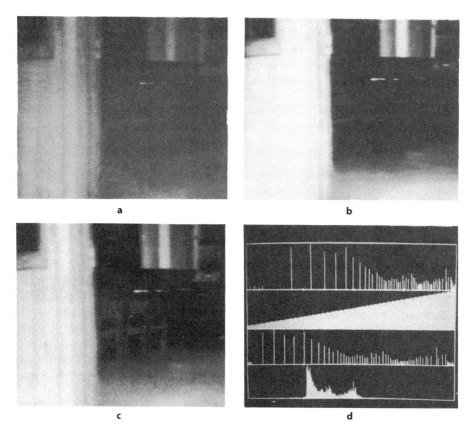

FIGURE 8.5 Illustration of the histogram specification method. (a) Original image (b) Histogram-equalised image (c) Image enhanced by histogram specification (d) Histogram *(from Gonzalez and Wintz (1987))*

If we were to set gray levels g_k ($k = 0, 1, \ldots, 7$) to zero when $k \leq 4$ our histogram consists only of the lower peak to which the corresponding image is:

0	0	0	0	0	0	0	0
0	0	0	0	0	0	0	0
0	0	7	5	6	6	0	0
0	0	6	5	5	7	0	0
0	0	7	7	7	6	0	0
0	0	6	6	6	5	0	0
0	0	0	0	0	0	0	0
0	0	0	0	0	0	0	0

FIGURE 8.6
The result of thresholding the gray levels of figure 8.1

The net result of this operation is that the gray levels which comprise the square in the image are enhanced relative to the background. On a graphics display of the image therefore, the square would generally appear as more distinct than in a display of the original image, due to the fact that levels constituting the background have been reduced relative to those constituting the square.

This technique of setting certain gray levels to zero relative to others is commonly called 'thresholding'. The effectiveness of the method generally depends upon the histogram of the gray level distribution on the original image exhibiting at least two identifiable peaks, so that at least one or other of the levels contributing to the peaks can be set to zero. In such cases the contrast in the image is generally enhanced.

8.5 IMAGE SMOOTHING

This approach renders the image more acceptable to the eye by reducing the contribution made by sharp transitions, or the occurrence of spurious peaks which may arise from poor sampling or errors occurring in image transmission. The techniques involved in smoothing may be based on spatial as well as frequency (Fourier) domains, but for the purpose of this chapter we present only the spatial approach.

8.5.1 Neighbourhood Averaging

Neighbourhood averaging is simply a process by which a pixel gray level may be replaced by the average of the gray levels of the surrounding pixels over a specified area.

Consider a subsection of an image represented by the set of gray levels shown below:

0	1	0	2	2	1	2	0
0	0	2	7	7	7	1	0
1	1	1	2	0	1	2	0

Notice the three gray levels corresponding to the value 7 which represents a sharp transition in levels in the image subsection by comparison to the other levels.

Defining a neighbourhood or area (sometimes called a window) around a specified gray level (denoted by O below) as the eight surrounding pixels (denoted by X), then our averaging domain can be represented as:

X	X	X
X	O	X
X	X	X

If we position the above window so that the O coincides with the first gray level with value 7 and then determine the average including all gray levels occurring at position X, the value of 7 would be replaced by the value:

$$(0 + 2 + 2 + 7 + 0 + 2 + 1 + 2)/8 = 2$$

If we now shift our window one pixel to the right and continue the process the next gray level value is determined as:

$$(2 + 2 + 1 + 7 + 1 + 0 + 2 + 7)/8 = 3 \text{ on digital rounding.}$$

Again with a further shift to the right our final value will be determined under the same process to be 3, so that on replacement of existing levels by those determined above our new image subsection becomes:

0	1	0	2	2	1	2	0
0	0	2	2	3	2	1	0
1	1	1	2	0	1	2	0

In other words, the sharp transition has been eliminated.

Mathematically the above process may be expressed as:

$$f'(x, y) = \frac{1}{N} \sum_{p,q \in S} f(p, q)$$

for $x, y = 0, 1, \ldots\ldots\ldots, N - 1$

In the above S represents the set of co-ordinates of points in the neighbourhood of the point (x, y), excluding the point x, y itself and N is the total number of points in the neighbourhood.

8.6 IMAGE SHARPENING

In the previous section we showed that by adding or averaging gray levels we can achieve a smoothing effect to the image. In this section we show, conversely, that by subtracting neighbouring gray levels we can achieve a sharpening or highlighting effect at different parts of the image.

This process of subtracting gray level values generates a further set of values commonly referred to as gradient values. Two approaches using subtraction methods are defined in relation to a pixel gray level at location x, y as follows:

(1) Form the absolute difference between the gray level at position x, y and adjacent levels at locations $x + 1, y$ and $x, y + 1$ then add the result;

(2) Form the absolute difference between the gray level at x, y with that at x + 1, y + 1 and add the result to the difference between gray levels at x, y + 1 and x + 1, y.

(1) and (2) above are represented diagramatically below:

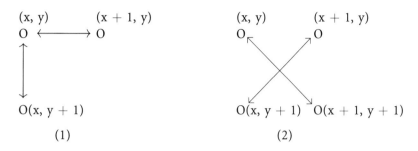

(1)   (2)

Mathematically, the operations can be expressed in terms of the gray levels g(x, y),g(x + 1, y) etc as:

(1) $G[g(x,y)] \approx |g(x,y) - g(x+1,y)| + |g(x,y) - g(x,y+1)|$

(2) $G[g(x,y)] \approx |g(x,y) - g(x+1,y+1)|$
$\quad + |g(x+1,y) - g(x,y+1)|$

In the above [G(x, y)] denotes the gradient and the approximation sign has been used as the values determined above are not strictly equal to the true gradient which mathematically is:

$$G[g(x,y)] = [(\partial g/\partial x)^2 + (\partial g/\partial y^2)]^{\frac{1}{2}}$$

However, for practical purposes the above approximation to the gradient is computationally faster whilst yielding acceptable results.

Having determined our gradients [G(x,y)] for all positions of x,y we may then form new images ie replace our original set of gray levels by the gradient values themselves, so forming two images for the set of g(x,y) where:

$$g(x,y) = G[g(x,y)]$$

for the two types of gradient defined above.

In the case of image borders where we have no pixels outside the actual border with which to specify the window, we simply repeat the gradient values determined at one pixel within the border. From what has been described above, it should be fairly obvious that for image regions, other than background, consisting of similar gray level values, the difference determined within the region will give gray levels of low magnitude and probably, therefore, more consistent with the levels forming the background. The effect of the operation at an edge say, or where we transfer from background to a region of higher gray level values, is to leave the edge values relatively unchanged. The method is therefore attractive as a way of highlighting edge regions in an image. (See figure 8.7 below).

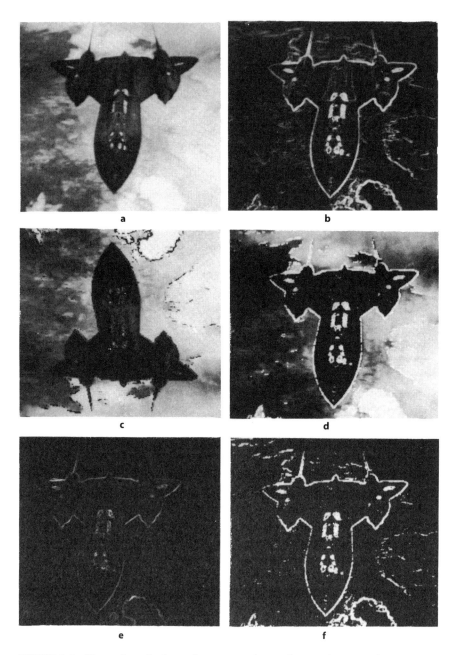

FIGURE 8.7   Illustration of edge enhancement by gradient techniques *(from Gonzales and Wintz (1987))*

9 Image Segmentation and Analysis

A. BARRETT

Image processing can be broadly classified into two main areas:

ENHANCEMENT AND ANALYSIS

The former chapters concerned with image processing presented a number of techniques directly related to the processes of enhancement.

This chapter is largely concerned with the presentation of techniques of analysis. For example, consider the image of Figure 7.1d, which shows the periodic structure within the image obtained through the Fourier filtering process.

This image, as presented to a scientist working in the field of virology, may cause questions to be asked about the extent and uniformity of the structure within the image.

Similarly, a meteorologist presented with a satellite image showing various weather formations may require further information about the particular size and distribution of specific cloud formations in relation to the area under consideration.

To assist in providing answers to such questions requires that we firstly further process the image in a way which renders it suitable for subsequent analysis. This form of processing requires that we initially identify and isolate the components of interest within the image.

The above process is generally referred to as SEGMENTATION. The segmentation approach is based on the detection of changes in the gray level distribution within the image. For example, a boundary can be considered as a line of pixels whose gray levels are distinctly different from the gray levels of the pixels either side of it. Detection of boundaries within an image can be achieved therefore, by identification of groups of pixels with this characteristic.

The early sections of the chapter explain the processes concerned with the detection of regions in an image which may be identified as single points, lines or boundaries. The latter sections explain how methods for representing the image may be applied to assist in any subsequent analysis. The chapter concludes with a section illustrating methods for the identification and analysis of image texture.

117

9.1 DETECTION OF POINTS, LINES AND EDGES

As pointed out above, a region in an image may be characterised by a significant change in the gray level distribution in a specified direction. For example, consider the single line of gray levels below:

$$0\ 0\ 0\ 0\ 0\ 0\ 0\ 7\ 0\ 0\ 0\ 0\ 0\ 0\ 0\ 0$$

Given the information that our background levels are represented by gray levels having a value zero, we may conclude that at the eighth pixel from the left we have a non background level and that this level may also characterise a single point. Basically the gray level at the eighth position represents a discontinuity in that all the other levels are of uniform value ie zero, except for the pixel at position eight. The process of recognising such discontinuities particularly when they occur in groups, may be extended to the detection of lines within an image.

Most line and point detection algorithms involve the positioning of a 'window' or matrix of values at successive pixel positions within the image and then forming products with the values in the matrix and gray levels at successive pixels.

A typical matrix is illustrated below and consists of a 3 by 3 set of values v_1 to v_9

$$\begin{array}{ccc} v_1 & v_2 & v_3 \\ v_4 & v_5 & v_6 \\ v_7 & v_8 & v_9 \end{array}$$

Initially the matrix is located to fit into the top left corner of the image with v_5 positioned at one pixel to the right and one pixel below the top leftmost pixel of the image. Particular values are ascribed to the elements $v_1,....., v_9$ (see below). If we denote the gray levels in the image corresponding to the matrix values $v_1,........8, v_9$ by $g_1,........8, g_9$ we start the detection process by forming the product:

$$P = v_1 g_1 + v_2 g_2 + ... + v_9 g_9$$

$$= \sum_{i=1}^{9} v_i g_i \tag{9.1}$$

We then shift our mask one pixel to the right and repeat the calculation. The process is repeated for all the pixels in the line (with the exception of the edge pixels). We then reposition the matrix one pixel down and at the left edge of the image and repeat the procedure. Note that borders are ignored, particularly as many images are of dimension 256 by 256 or more and the effect of ignoring the outside edge is generally insignificant, for images of these dimensions.

At the end of the entire process we have a set of P values from which we are able to determine the existence of a line. The basic rule for establishing the existence of a line is whether among the set of P values subsets of P values can be found for which:

$p \geq T$ (where T corresponds to some arbitrary value).

Below are shown a set of matrices for detecting lines oriented horizontally, vertically and at $^{+}_{-}45°$

-1	-1	-1	-1	-2	-1	-1	-1	2	2	-1	-1
2	2	2	-1	2	-1	-1	2	-1	-1	2	-1
-1	-1	-1	-1	2	-1	2	-1	-1	-1	-1	2

Horizontal	Vertical	+45°	−45%
Detector	Detector	Detector	Detector

9.1.1 Detection of Isolated Points

The process for the detection of isolated points is very much the same as that described for lines except that the matrix values are assigned a different set of values to those assigned for the detection of lines. One favoured set of values is shown below:

-1	-1	-1
-1	8	-1
-1	-1	-1

To illustrate the technique consider the distribution of gray levels below:

0	1	1	1
1	6	0	1
0	0	2	0

Positioning our matrix at the top left would give for the first evaluation of P in accordance with equation 9.1 ie forming products and adding:

$$P = (-1)^{*}0 + (-1)^{*}1 + (-1)^{*}1 + (-1)^{*}1 + (6)^{*}8 + (-1)^{*}0$$
$$+ (-1)^{*}0 + (-1)^{*}0 + (-1)^{*}2 = 43$$

Shifting the matrix one pixel along yields for P:

$$P = (-1)^{*}1 + (-1)^{*}1 + (-1)^{*}1 + (-1)^{*}6 + (0)^{*}8$$
$$+ (-1)^{*}1 + (-1)^{*}0 + (-1)^{*}2 + (-1)^{*}0 = -12$$

In general we use |P| and say that a single point has been detected if $|P| \geq T$ where T, as pointed out in the previous section, is some arbitrary value selected on the basis of the a-prior knowledge of the image application area.

As can be seen from the example above, the value determined for $|P|$ in the image is significantly higher at the position corresponding to the highest gray level value, consistent with a level constituting a point.

9.1.2 The Hough Transform

This transformation enables the automatic detection of lines, straight or curved, in any orientation. The problem of detection may be formulated as that of locating all the points in an image which identify with straight lines or curves.

Mathematically, the problem of locating lines given a set of points is a fairly simple one although standard approaches become computationally prohibitive for all but the simplest images.

An alternative approach, proposed by Hough (1962) has the attraction of significantly reducing the computational overheads incurred by the conventional approach.

The equation for the straight line connecting a set of pixels is given by the formula:

$$y = mx + c \qquad (9.2)$$

where m corresponds to the slope of the line and c its intercept with the y axis.

The approach adopted by Hough was to reformulate the above equation as:

$$c = y_i - mx_i \qquad (9.3)$$

and to restate the problem in the following way:

Given a specific value for m and a set of co-ordinate pairs x_i, y_i determine the corresponding value for c from equation 9.3.

We can then say that for those x_i, y_i and a specified m which yield the same values for c that we have located the line (slope m, intercept c) associated with those points. The problem with the method is that for lines either vertical or tending towards the vertical, the intercept value for c tends to infinity which clearly leads to computational problems. To overcome this the normal representation of the line is used:

$$x \cos \theta + y \sin \theta = \rho \qquad (9.4)$$

The problem may now be regarded as the determination of ρ for a specified value of θ from the x_i, y_i values inserted into equation 9.4.

In practice however, we are generally more concerned with the identification of points that are associated with particular lines. The co-ordinate pairs together with the specified values when inserted into equation 9.4 will generate a range of values for ρ. We may then say that

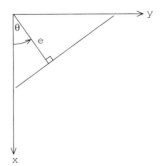

FIGURE 9.1
Normal representation of the line

those values of ρ which lie within a specified radius or neighbourhood are associated with the line defined by the ρ, θ values determined by the point at the centre of the neighbourhood.

An example is given to illustrate the process of the Hough Transform in relation to two points, however it should be noted before studying the example that a property of the Transform is to map single points with x, y co-ordinates into sinusoidal curves in ρ, θ space.

For example, we have chosen a situation represented by two points A, B with respective co-ordinate values (1,1) ; (2,2) as shown in figure 9.2.

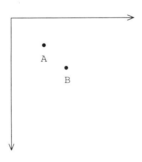

FIGURE 9.2
Two points A, B to be mapped by the Hough transform

Suppose we select θ_{min} as $-90°$ and θ_{max} as $+90°$ and decide to increment θ in steps of $+45°$. Then for each of the points A,B we can establish the following table for the values defined by equation 9.4.

TABLE 9.1   Values of the mapping under the Hough transform for the two points A and B

For the point A:	θ	$x\cos\theta$	$y\sin\theta$	ρ
	$-90°$	0	-1	-1
	$-45°$	$1/\sqrt{2}$	$-1\sqrt{2}$	0
	$0°$	1	0	1
	$+45°$	$1/\sqrt{2}$	$1/\sqrt{2}$	$\sqrt{2}$
	$+90°$	0	1	1

TABLE 9.1 *(cont.)*

For the point B:	θ	$x\cos\theta$	$y\sin\theta$	ρ
	$-90°$	0	-2	-2
	$-45°$	$2/\sqrt{2}$	$-2/\sqrt{2}$	0
	$0°$	2	0	2
	$+45°$	$2/\sqrt{2}$	$2/\sqrt{2}$	$2\sqrt{2}$
	$+90°$	0	2	2

A corresponding plot of θ.vs ρ shows the mapping of the points as determined by Equation 9.4.

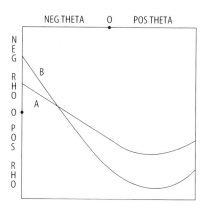

FIGURE 9.3
Mapping of the points A,B under the Hough transform

The results may be interpreted as follows:

We firstly observe that for the values corresponding to $\rho = 0$ and $\theta = 45°$, the two curves intersect. The point of intersection identifies that the two curves (or points in x, y co-ordinate space) are common to the points having a line of slope $-45°$ with perpendicular bisector to the line having a slope of zero.

This result is clearly confirmed by the line joining the two points A and B.

The method may be extended to the detection of points associated with analytically defined curves and the reader is referred to the further reading material for details of the techniques.

9.2 IMAGE TEXTURE

The texture of an image can be defined in terms of its smoothness or its coarseness sometimes called graininess. One area of image processing in

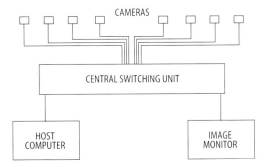

FIGURE 9.4   Basic components in a vision system

which the quantification of texture, as defined above, plays a crucial role is that of industrial vision.

These systems may be used to assess the quality of products by measuring the texture of the surface of the product. In general a poor surface, such as one showing a high degree of graininess, implies a poor product. The basis of industrial vision systems is on the use of intelligent cameras and appropriate software for image analysis. Figure 9.4 illustrates such a system.

The attraction in using an industrial vision system of the above type is that they are more sensitive than manual approaches to the identification of surface defects. Furthermore, manual inspection systems are subject to varying standards and involve fairly high costs. These factors together with the degree of subjectivity involved in using a manual approach makes the vision systems an attractive alternative.

The remainder of this chapter is devoted to explaining two methods for the identification and assessment of texture for a specified region. The methods are based on:

1. The statistical properties of an image; and
2. The Spectral or Fourier characteristics of the image.

The use of statistical techniques enables the characterisation of the texture in terms of smoothness or coarseness whereas the spectral approach is based on the properties of the Fourier spectrum (Chapter 7) and is used to detect regular or periodic features which may be present in the image. Examples showing images exhibiting different textural features are shown in figure 9.5.

9.2.1 Statistical Methods for Texture Description

For an image with a corresponding histogram distribution of gray levels $p(z_i)$ where z denotes the i'th gray level and $p(z_i)$ its frequency, the n'th

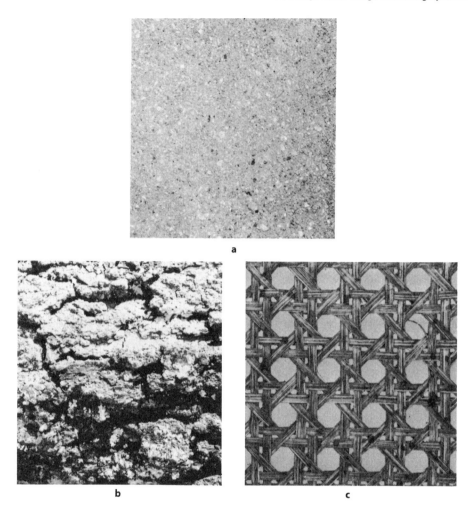

FIGURE 9.5 Examples of (a) smooth; (b) course; and (c) regular textures *(from Gonzalez and Wintz (1987))*

moment of z (the random variable denoting gray levels) about the mean can be written as:

$$\mu_n(z) = \sum_{i=1}^{L} (z_i - m)^n p(z_i) \tag{9.5}$$

Where L is the number of distinct levels.

m is the mean value of z ie the average intensity:

$$m = \sum_{i=1}^{L} z_i\, p(z_i)$$

Note that $\mu_0 = 1$ and $\mu_1 = 0$ from equation 9.5.

The second moment, sometimes called the variance $\sigma^2(z)$ determines gray-level contrast and may be used to establish descriptions of relative smoothness. In particular the value for R in equation 9.6 is zero for areas of constant intensity and approaches unity for large values of $\sigma^2(z)$

$$R = 1 - \left(1/\left(1/ + \sigma^2(z)\right)\right)$$

Additional information such as the skewness of the histogram can be obtained from the third moment whereas the fourth moment provides details on the relative flatness.

The fifth and higher moments do not provide much information about histogram shape but do provide further quantitative discrimination of texture content (Gonzalez and Woods (1992)).

9.2.2 Fourier or Spectral Methods for Texture Description

As shown in Chapter 7, determination of the Fourier spectrum provides a useful way of identifying any periodic or regular features within an image.

Three features of the spectrum that have been found useful for texture description are:

- the significant peaks in the spectrum give the principal direction of the texture patterns;
- the location of the peaks in frequency space provide the spatial period of the patterns; and
- filtering of periodic components leaves non-periodic image elements which may then be described by statistical techniques.

In fact this latter step corresponds to the converse filtering operation used to generate the biological image in Chapter 7 in which the non-periodic components of the spectrum were filtered out prior to involving the inverse transform operation.

The approach to abstracting and analysing the features in the spectrum is to express the spectrum in polar co-ordinates to yield a function $S(r, \theta)$, where S is the spectrum function and r and θ are the variables in the co-ordinate system. For each direction θ and each frequency r, both $S(r, \theta)$ and $S_\theta(r)$ are considered to be one dimensional functions. $S_\theta(r)$ for a fixed value of θ yields spectral characteristics along a radial direction from the origin, whereas $S_r(\theta)$ for a fixed value of r yields the behaviour along a circle centred on the origin (Gonzalez and Woods (1992)).

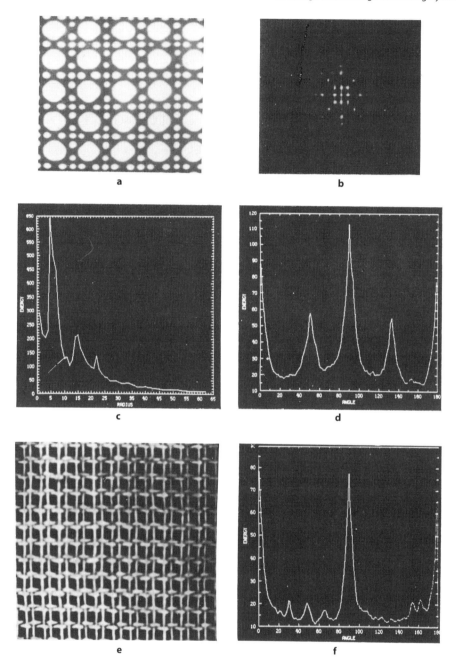

FIGURE 9.6   (a) Image showing periodic texture. (b) Spectrum. (c) Plot of S(r). (d) Plot of s(θ). (e) Another image with a different type of periodic texture. (f) Plot of s(θ) *(from Gonzalez and Wintz (1987))*

A more global description is obtained using the functions:

$$S(r) = \sum_{\theta=0}^{\pi} S_{\theta}(r)$$

$$S(\theta) = \sum_{r=1}^{R} S_r(\theta)$$

where R is the radius of a circle centred at the origin. Figure 9.6 illustrates the use of the above equations in relation to image texture for images with different periodicities.

Part 3
Knowledge-Based Systems and Image Processing

D. GRAHAM and A. BARRETT

10 Introduction

D. GRAHAM and A. BARRETT

This part of the book aims to provide a synthesis of the two areas of knowledge-based systems (kbs) and image processing systems, plus hybrid systems for the biomedical field. It describes the investigation and analysis of current work, of methods used, techniques and applications. Forming conclusions on the degree of progress made and outlines future recommendations.

10.1 FUNDAMENTAL STAGES OF KNOWLEDGE-BASED SYSTEMS DEVELOPMENT AND IMAGE PROCESSING

An overview of knowledge-based systems has been provided in Part I. The main stages in carrying out knowledge-based systems development are acquisition, representation, development and inferencing mechanisms. Trends and current research issues in these areas have been identified, as well as a number of limitations or problems inhibiting knowledge-based system development. However, knowledge-based systems and knowledge engineering are now well established and cover diverse applications in fields such as medicine.

An overview of image processing has been provided in Part II. The main stages in carrying out image processing are image acquisition, pre-processing, segmentation, representation and description, and recognition and interpretation. Main techniques involved include Thresholding and the Fourier Transform, which have been outlined. Texture analysis with respect to medical applications, is also an important method in image processing, since it can allow for identifying characteristics in biomedical images.

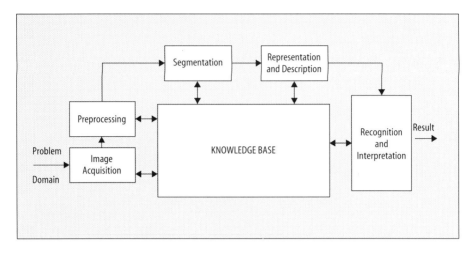

FIGURE 10.1   Fundamental Steps in Image Processing   *(Gonzales and Woods, 1993, p. 8)*

10.2 LINKING BASIC STEPS IN IMAGE PROCESSING AND KNOWLEDGE-BASED SYSTEMS

The basic steps involved in image processing are defined by Gonzalez and Woods (1992). Beginning with image *acquisition*, where a digital image is acquired through means such as a scanner or television camera, producing a two-dimensional image. The next step is *pre-processing* which aims to improve the quality by enhancing contrast, removing noise and isolating regions of alphanumeric information. Thirdly, we have *segmentation* where the image is partitioned into constituent parts or regions. Data can be represented as a boundary or region. Boundary representation concerns external shape characteristics, whilst regional representation focuses on internal characteristics such as texture or skeletal shape. *Description*, or feature selection involves feature extraction, allowing for highlighting of features of interest. *Recognition* labels an object from the descriptors whilst the final step, *interpretation*, assigns meaning to the recognised objects. A knowledge base is depicted, where knowledge concerning a problem domain is kept in a knowledge database and encoded into the image processing system. Interaction occurs with the knowledge base in the steps due to a perceived prior knowledge on what the result will be. This process is shown in figure 10.1 (Gonzalez and Woods, 1992 p. 8). This diagram however has since been considered inexact in terms of KBS interaction, the extensive links to the term knowledge-base when knowledge-based system or database system appears to be more the intent of the authors, for instance. We can now consider in more detail the techniques and relevant areas.

11 Current Knowledge-Based Image Processing Systems

D. GRAHAM and A. BARRETT

11.1 KNOWLEDGE-BASED SYSTEMS AND IMAGE PROCESSING

The two fields of knowledge-based systems and image processing are described: links, current combined use in applications and what various authors have to say.

11.1.1 Basic Concepts

Image processing is an aspect of the computer vision area. The approach by knowledge-based systems to the image processing steps is classified in terms of low-level and high-level vision (Rao and Jain, 1988) or low, intermediate and high-level processing by Patterson (1990).

Low-level vision is based on the extraction of features, resulting in a segmented image with labelled different regions. Shapes, spatial interrelationships, and surfaces of objects may be described. *High-level vision* attempts to consistently interpret labels obtained from low-level processing using a 'priori information about the scene's domain'.

The two main steps can thus be considered as segmentation and interpretation. The area of AI has shown an interest in the concept of image processing, but mainly to image understanding or interpretation. When distinguishing between the two terms, Samadani (1993) states *image understanding* to have intermediate and final results in symbolic form as opposed to numeric in *image processing*. Image understanding also has lists containing world knowledge, or knowledge on methods for processing images, whilst interpretation is a mapping from sensory data to the model, where domain relevant concepts are represented (Rao and Jain, 1988).

11.1.2 Image Understanding and Image Processing Problems

Pinker (Firebaugh, 1989) defines problems in automating processes of understanding. These are in terms of shape recognition in template matching, feature extraction, Fourier analysis and structural descriptions. To overcome these problems and general image understanding problems, David Marr's (1982) approach in Patterson (1990) to successful systems, can be realised by using a representational scheme in each of the following steps. (This can aid in shape and image understanding.) The steps are:

(1) Gray-scale image – pixels are used, with representation to aid local and first-order statistical transformations;
(2) Raw Primal sketch – this is a two-dimensional sketch to ensure objects are explicit. Edge segments, and features such as texture, are present. Pictorial image descriptions, however, require more emphasis;
(3) Two and a half-dimensional sketch – emphasis is required on the depth of surfaces, or any discontinuities, shape, and texture;
(4) Three-dimensional models – representations are symbolic providing relational, geometric and attribute description. Cones and cylinders may aid in representing object types.

Marr's suggestion can be dealt with (Firebaugh, 1989), but the latter step causes problems because parallel architectures may be required. These have been present in a number of recent applications and articles. More recent problems to address in image understanding and image processing are stated by Samadani (1993). These problems are currently inhibiting image processing development and subsequent research and applications. They are firstly, the fact of the volume of data required in performing a task, particularly in the construction of 3-D reconstructions. Secondly, the fact that dealing with spatial data requires knowledge of local properties, and finally, the requirement that numeric and symbolic computations are needed.

11.1.3 Issues to Consider

When viewing how knowledge-based systems may be used to aid in image processing, we can consider the following points:

Goal orientation

Stages in image understanding or image processing need to be goal oriented so that a particular achievement is met. For example, as Samadani (1993) states, a goal may be to represent image data in digital format to allow for easier storage, transmission, and processing. It may be to improve the image quality for the user, extract quantitative or qualitative information such as sizes, textures and shapes of objects, or to remove physical distortion during image acquisition.

Representation

We need to consider how knowledge can be represented. This could be in the form of predicate logic, semantic nets, frames or production rules (Tjahjadi and Henson, 1989). Rao and Jain (1988) discuss the advantages of using schemata or frames in image interpretation. They also state that knowledge representation causes issues of declarative or procedural adequacy to be considered. Declarative structures ensure that knowledge is explicit and allow for modification. Procedural adequacy is where procedures correlated to schemata may control interpretation.

Uncertainty

Problems of noise and distortion may mean uncertainty when an image is transformed on a computer. Methods such as Dempster-Shafer theory may be required to overcome evidential information.

Control Strategies

Rao and Jain (1988) state that here it needs to be considered whether parallel or sequential control need to be used. They favour sequential and global control as opposed to local, due to image understanding being sequential and cyclical. A decision is required to be made on using top-down or bottom-up control. They favour bottom-up for segmentation and top-down for the rest of the process. Another issue is to decide whether control is to be distributed or centralised.

Linking

The way the link is to be made between the two areas is discussed in the next section.

Memory

A distinction needs to be made in terms of short-term or long-term memory. Long-term memory can be used to hold priori knowledge, whilst short-term memory is the internal model or description within the image.

User interface

Many articles now discuss the use of expert or users in development of the knowledge-based image processing system. (See below for example).

11.2 CORRELATING KNOWLEDGE-BASED SYSTEMS AND IMAGE PROCESSING

Rao and Jain (1988) argue that image interpretation may be too difficult due to uncertainties in formulating hypotheses on the data or model. They go on to distinguish between two types of approaches using knowledge in image processing proposed by Browse (1982). These are the *image-feature-access* approach and the *volume-access* approach. In the first, model knowledge of objects is used to match image features against features predicted in the models. But it is argued however, that difficulty will arise in interpretation before segmentation and hence some form of iteration may be required. The volume access approach uses modules with context-free representations.

11.2.1 Knowledge Types

The types of knowledge that can be used to aid in image processing, are described by Plant *et al.* (1989) and can be classified as follows:

- Knowledge of the picture entity and its environment – assumptions can be made for knowledge on entities or general world knowledge;
- Knowledge about the construction of the picture – knowledge of methods and tools used;
- Knowledge about the picture as a representation – knowledge representation schemes may be used for object representation and description, which tend to be application specific. Knowledge on the object domain may also be used;
- Knowledge about human interpretation of pictures – based on analysis of picture interpretation;
- Knowledge of human perception.

Prior knowledge is used, which tends to be more application specific at higher levels of image processing such as understanding and recognition. The next aspect illustrates a specific framework for combining knowledge-based systems and image processing and goes on to consider the general coupling of image operations with knowledge-based systems and processes as mentioned in section 11.1.3.

This section presents an example of an approach to the combination of knowledge-based systems and image processing in an attempt to interpret a certain class of images. The example is from Walker and Fox (1987) in which they pose the question:

How can expert systems be used to help in the interpretation of medical images?

In this instance the images refer to brain scans acquired for the purposes of possible tumour identification. An example of a brain scan in which the tumour can be clearly seen has already been illustrated in Chapter 6. However this particular scan is not necessarily representative of all brain images and the problem of disease identification is not always obvious.

Walker and Fox (1987, pp. 253–254) begin their analysis by stating the following proposition or fact. (The example is in pseudo-English, the representation language of Props 2, a general knowledge engineering package developed in the authors' laboratory. Capitalised strings are logical variables which are universally instantiated during inference):

'Given the following proposition or fact:
   symptoms of cerebella tumour include dizziness
and the patient data:
   history of J Smith includes dizziness
the following rule:
   if history of Patient includes Symptom
      and symptoms of Disease include Symptom
      and Disease of Patient is not eliminated
   then possible diagnosis of Patient include Disease
will generate the following conclusion:
   possible diagnosis of J Smith include cerebella tumour
Given a set of possible diagnoses generated from patient data we can now try to verify hypotheses on the image data. This is the classic generate-and-test strategy, and is probably the commonest approach to signal processing in AI. It can be conveniently illustrated using control-rules to specify a strategy for processing, encoding or searching images in ways which depend upon the hypothesis which is being pursued (or to reflect other goals). Given the above conclusion and the following facts:

● common locations of cerebella tumour include posterior fossa
● morphologies of cerebella tumour include dense sphere
   Then this rule will be executed:
      if possible diagnoses of Patient include Disease
         and common sites of Disease include Site
         and morphologies of Disease include Morphology
      then verify presence of Morphology at Site
   the following goal is generated:
      verify presence of dense sphere at posterior fossa

In passing we might note that the examples are written in a **general** form; they are not specific to particular domains, particular medical conditions or imaging devices. Consequently as more medical facts are added to the knowledge base the scope of application of the rule set increases, but without making potentially troublesome alterations to the rule base and hence the logic of the inference strategy. Many expert systems depend upon knowledge bases which are built entirely out of special case rules; this limits the scope for identifying general strategic knowledge. In image interpretation as in other areas we expect a general trend towards representing knowledge in generalised logical form because of the power, flexibility, and scope for formal analysis which this offers.
   Many knowledge sources could guide the generate and test process. For example image analysis could yield some features which are easily interpreted in light of clinical information, while others are not easily explained unless other findings could be confirmed on the image. The following rule is another strategic knowledge source:

   If unexplained features include Feature
      and Feature is present at Site
      and Feature is possible part of Morphology
   then verify presence of Morphology at Site

We now come full circle and generate a set of possible interpretations from what we have discovered about the image:

> *If Morphology is present at Site*
> *and morphologies of Disease include Morphology*
> *and current patient is Patient*
> *and Disease of Patient is not eliminated*
> *then possible diagnosis of Patient include Disease*
> *and site of Disease is Site*

An outstanding question concerns the recognition of objects and events; what does it mean, for example, to 'verify' a hypothesis on an image?'

The example quoted depends upon the precise definition of the term 'verify' ie confirm or otherwise a specific hypothesis on an image.

To achieve this requires the coupling of image processing to image interpretation. This necessarily requires that the steps of image segmentation, such as those described in chapter 9, are invoked to abstract specific features or structures form the pixel array before presenting them to the expert system.

11.2.2 Five Methods of Linking Image Processing and Knowledge-Based Systems through Image Interpretation

(a) *Quantitative coupling* – Image processing algorithms are applied to the pixel array, giving singleton parameters, distributions and derived Boolean descriptors such as normal and abnormal. The data can then be passed for interpretation by the knowledge-based system. The authors do, however, argue that such a system may not be classified as image understanding because there is no explanations of conclusions derived;

(b) *Human Encoding of Image Data* – Since humans may be able to identify shapes and objects, a knowledge-based system may be developed to interact with a human viewing the imaging output. Benefits of using this scheme are that it is simple to implement but may require costly resource – the expert or skilled person;

(c) *Automatic symbolic encoding of the image* – Symbolic description is described, which is defined as interpreted explicit image representation, with observed parts explained and implicit knowledge made explicit. Descriptions from components such as edges can be used to identify structures in the image considered to be interpreted correctly, to further interpret ambiguous structures. The authors define automatic symbolic description as a difficult process. In medicine, however, advances are making this easier such as in medical radiology;

(d) *The Image as a Database* – Due to problems with there being more than one consistent interpretation of images and descriptions not being absolute, an image description may be considered as a

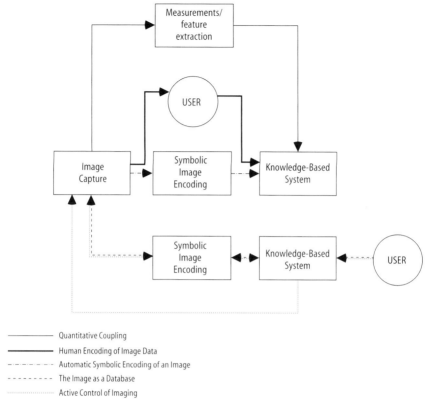

FIGURE 11.1 Five Methods of Linking Image Processing and Knowledge-Based Systems through Image Interpretation *(Walker and Fox, 1987 p. 255)*

database of descriptions which the knowledge-based system can search and add to as new descriptions are derived. Knowledge may be encoded so that a query may be satisfied;

(e) *Active control of imaging* – Knowledge-based systems may be required to control image capture and low level processing. Thus adjustments could be made in terms of system goals or prior image analysis. A knowledge-based system would require domain knowledge on significance of certain structures, and imaging knowledge of 'know-how' of the image itself. This is shown in figure 11.2 below.

11.3 APPLICATIONS

Before classifying in detail the applications of knowledge-based systems and image processing, we can consider an application that shows knowledge-based image processing steps, fitting in the concepts

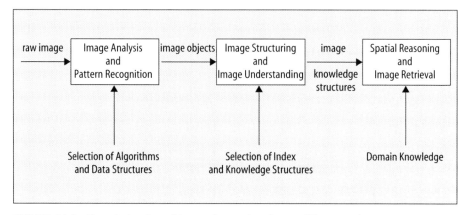

FIGURE 11.2   Knowledge-Based Image Processing Stages *(Chang and Hsu, 1992, p. 432)*

considered so far. Chang and Hsu (1992) state that in the image information systems, the image processing and image database are required to perform three stages, which they classify as knowledge-based image processing. These are shown in figure 11.3.

These steps are described as follows:

(a)   *Image analysis and Pattern Recognition* – The raw image is analysed with recognition of image objects. The image objects are encoded in the data structures such as quad-trees (see Gonzales and Woods (1992), for further explanation) to allow for further handling. Image objects can be classified as objects with various attributes such as co-ordinates. Unanalysed parts of the image may be later processed. Image processing algorithms and data structures are the system input in this first stage;

(b)   *Image Structuring and Understanding* – For a number of applications, the image objects are required to be converted into image knowledge structures, to support spatial reasoning and information retrieval. Image knowledge structure will depend on application domain knowledge and features to be indexed. Image knowledge structures are inputs which can take a number of forms such as semantic networks or directed graphs of spatial relations;

(c)   *Spatial Reasoning and Image Information Retrieval* – Certain applications require spatial reasoning or image information retrieval. Problems in a domain may be solved by a domain knowledge base. Transformations may be carried out on the image knowledge structure. The result from this stage is user-specific knowledge structure.

We can now consider the areas where knowledge-based systems and image processing have been used, considering early applications, present and future possibilities.

11.3.1 Early Applications

VISIONS was created for interpretation of outdoor colour scenes from identifying regions. SIGMA was designed to aid in aerial photographs in searching for road and houses. ACONYM was designed to be used to recognise aircraft images, but was developed as a general purpose vision system representing objects and parts (Brookes *et al.*, 1979). It is one of the most successful early systems in vision. It consists of four parts. These are modelling, prediction, description and interpretation. It is a domain independent system, which allows users to specify world knowledge in the modelling phase. Interpretation is done by matching a picture and prediction graph. Object descriptions are within an object graph, with component relationships in a restriction graph. General geometric reasoning is one of its key features.

11.3.2 Fields of Impact

Image processing and knowledge-based systems have been widespread in fields such as the military, robotics, manufacturing and satellite image interpretation. One of the biggest uses has been in the medical profession, as discussed in chapter 13. We will now consider applications which have used knowledge-based systems, and their concepts with any special issues also addressed.

11.3.3 General

Gillies and Hart (1988) use knowledge-based systems methods and tools to restore an image in photo elastic stress analysis. They discuss using an expert to help with providing knowledge for restoration. Due to skill differences between novices and experts, the article discusses modelling some of these problem-solving strategies such as rules in a computer system. They use Pascal for programming on an APPOLLO DN3000. The authors argue that explanation is not considered necessary, and quality aspects need further consideration.

Tjahjadi and Henson (1989) discuss the uses of knowledge-based systems in image processing for interpreting line drawings (Shapira, 1984), and for image understanding (Lowe, 1987) in identifying features consistent in transformation of objects in 2-D images. They create a system using an expert system shell (XI plus) using some of these methods for image understanding, with three knowledge bases. One is pattern recognition and feature extracting, the second a descriptor, and a third for knowledge retrieval and learning.

A knowledge-based system for analysing complicated fringe benefits is suggested by Kujawinska (1993). Matsuyama (1992) considers an image processing knowledge-based system using declarative knowledge for image recognition and understanding. Grimm and Burke (1993) consider an image processing knowledge-based system based on SPIDER (Tamura *et al.*, 1983), with its modular knowledge base, which can be used as a general approach for solving image processing problems, by being easily adapted into image processing applications.

As an example of optical image processing analysis, a knowledge-based system is discussed by Prabhakaran *et al.* (1990) for interpretation of human signatures in identification of a person using scanners. Zahzah *et al.* (1992) suggest an automatic satellite image process using a fuzzy connectionist image processing system, using neural networks to overcome problems requiring expert knowledge in image classification.

11.3.4 Geographical Uses

Also discussed are articles from the SPIE conference 1987 (Gillies and Hart, 1988) where a query processor in LISP is combined with FORTRAN image processing routines, for map interpretation (Alemany and Kasturi, 1988). Shu and Freeman (1988) use IF-THEN rules and forward-chaining for image segmentation. Gillies and Hart (1988) state that these systems do not fully use knowledge-based methods or automate, but model some human analysis using rules. The author discusses Yuichi Ohta of Koyota University's vision system for region analysis for colour scenes outdoors, such as bushes and trees. Colour is used in the segmentation process, with knowledge modelled as production rules. The rules are top-down and bottom-up processes, with weights used for uncertainty. Patterson (1990) states that this system reflects dealing with complex scenes with objects comprising of substructures, but does argue that the state of play is far from interpreting complex scenes.

11.3.5 Remote Sensing

Goodenough *et al.* (1994) discuss the Canada Centre for Remote Sensing (CCRS) creating an expert system shell for remote sensing data, to apply knowledge for programs involved with difficult image processing problems such as geographic information system updates. They consider an Interactive Task Interface (ILTI) which uses a knowledge-based system with a PROLOG module to answer image analysis queries. An ILTI has been created, and so a planner using a knowledge base is used to modify increment changes between an expert and image analysis dialogue. This ensures new knowledge-based systems are thus easier and quicker. Gilliot *et al.* (1993) describe a knowledge-based system, NEXSYS, for detecting and interpreting thin linear networks in remote sensing images. They use object-oriented rules and frames on a Sun Sparc IPC 4/40 in LISP.

12 Knowledge-Based Image Processing: The Way Forward

D. GRAHAM and A. BARRETT

12.1 THE IMPACT OF KNOWLEDGE-BASED SYSTEMS ON IMAGE PROCESSING

Many articles sate the benefits of using knowledge-based systems for image processing. Walker and Fox (1987) state that the two fields have developed separately with more focus being made on applications combining the two fields. Gillies and Hart (1988) state that knowledge-based methods applied to many domains, have been minimal in image processing. They argue that the field has a lot of scope for the image processing domain. However, a number of articles outline the increase in use of knowledge-based systems and methods for image processing whether it is for restoration of images, or improving the performance of a system. Image processing is often seen in work on AI vision with sections in AI books on computer vision or visual image understanding. More recent articles tend not to consider the degree of linkage between the two fields of image processing and knowledge-based systems. Instead they deal more on how improvements or image processing problems may be addressed using knowledge-based systems.

12.2 CURRENT RESEARCH ISSUES

Garcia (1991) states that research is based around using visual knowledge for high-level processing, either as direct models of the domain, priori information on the environment, or human expertise. High level processing to influence low-level is discussed, with issues on representation, control strategy and reasoning. Chang and Hsu (1992) state that the following are issues that need to be considered in this combined area of applications, citing some authors. If we consider the image as a database as suggested in section 11.2.2, then these issues can be considered.

Query Language and User Interface

Queries such as SQL or QBE, Query By Example, or QPE, Query by Pictorial-Example, are used in applications. A QBE approach ought to be possible to be used with querying images in 2-D or 3-D (Bimbo *et al.*, 1992). It is argued that there should be possibilities for the user to switch between paradigms, with tabular approaches for simplified queries. The user interface ought to support retrieval. The authors also ask the following questions that need to be addressed. How does one design a single point of contact (SPOC) workstation for multiple paradigm queries, decompression, data compression? How can a unified interface supporting multiple paradigm interaction and multiple visualisation modalities be designed?

Data Models

Data models can be defined as mathematical defined concepts expressing static and dynamic properties of intensive data applications. Data models may be in a number of forms, such as hierarchical, network, and relational. Image information cannot be accurately described, underlying data representations need to be hidden. Issues raised are how mapping is to be done, one-many or many-one, consistency issues, equivalence issues in multiple representations, and spatial entities (objects) and relationships (image features) within images, which have no semantic meaning. Thus, without semantic meanings associated to these, interpretations will vary amongst users. Automation problems are research concerns for mapping from semantic to image features. If queries are to be done by pictures, then information to go into the knowledge base or data model needs to be distinguished. Recent uses have been made of object-oriented approaches (Gupta and Harowitz, 1991).

Indexing Techniques and Data Structures

Problems to consider concern index representation, organisation and extraction with issues based on structuring image data and acquiring and selecting image indexes (features).

12.3 Summary

Knowledge-based systems are primarily used for interpretation and understanding of images. The type of knowledge that can be used has been discussed, along with issues to address when linking the knowledge-based system to image processing. Five possible suggestions of using a knowledge-based system and its techniques have been provided in Chapter 11. These need to be worked on to allow for more advancements. In applications there has been a movement from algebraic to heuristic methods with incorporation of technologies such as neural nets. Recently issues such as errors are being addressed, with proposals such as having control errors using control strategies, to allow for higher performance. Other topics are generalising applications and allowing for full automation.

13 Summary and Conclusions

D. GRAHAM and A. BARRETT

13.1 KNOWLEDGE-BASED SYSTEMS

Knowledge-based systems have had a big impact on industry and society. They have got past the stage of prototypes and are now in the application stage where they are being used for everyday work. We can consider the areas which are the most relevant. The 'knowledge acquisition bottleneck' needs to be fully overcome by making more use of automated tools, to ensure higher efficiency. More recent technologies need to be incorporated into the process such as group decision support systems. Knowledge can be represented for different control in systems such as blackboard systems. Systems can be implemented more quickly due to modularity.

The movement to larger knowledge-based systems has resulted in necessities of validation and verification, concepts taken from the software engineering field. This will ensure more belief in the systems and their concepts. Knowledge-based systems and their capabilities are being taken seriously in this way, since most ordinary programs and databases are also designed with reference to validation and verification standards. There has been a development of various shells and tools to build systems. Combinations of languages such as C and knowledge-based languages or techniques are helping to overcome previous limitations. It is likely that future use of these types of techniques will overcome the problems, especially problems of uncertainty. It is also necessary for future requirements to ensure knowledge-based system analogy, thus allowing for generalisation on knowledge to deal with problem solving and uncertainty.

Some examples of applications of knowledge-based systems with the types of control and representation strategies typically used have been described in Part I. It can be seen that the future is moving towards applications suited to industrial needs with aspects of new technology such as hypermedia in DentLE. Applications are also using fuzzy techniques to overcome or deal with problems of uncertainty.

13.2 KNOWLEDGE-BASED SYSTEMS AND IMAGE PROCESSING

The use of knowledge-based systems for image processing has been known since the 1970s, and has been widely used particularly in the early 1980s. This was found when searching for articles (Basra, 1995), where many references were dated to that time period. Knowledge has been used in applications, in the form of 'priori' for techniques such as segmentation. A knowledge-based system has also been used for interpreting and understanding images. The way the link is made between image processing and knowledge-based systems has also been reflected in a number of ways, such as using human computer interaction or automatic symbolic encoding. Many techniques use pattern matching. There have been movements to more modular knowledge-bases, and many techniques such as frames, and objects are being used. It is often seen that combinations of languages are used, with some languages such as LISP.

   In terms of considering links between the two fields, very few articles discuss bringing together the two fields. Most articles just generally state that KBS ideas and techniques can be used to aid images in ways such as diagnosis or interpretation of images, by allowing for recognition and pattern matching. Many systems such as SUN SPARC are currently being used, allowing for more flexibility. Numerous knowledge-based systems and image processing prototypes exist still at the experimentation stage with comparatively few as fully automated applications, these are mainly for remote or geographical uses.

13.3 KNOWLEDGE-BASED SYSTEMS AND IMAGE PROCESSING IN THE BIOMEDICAL DOMAIN

MRI and CT techniques are often used to obtain images, with PET (Positron Emission Tomography). As in the previous section SUN SPARCS are being used with programming languages such as C. Shells such as Leonardo are also being used. Previously, there were a lack of tools and shells for building these systems which also meant a number of limitations. The problem with biomedical images is that a Human Computer Interface (HCI) is still needed to recognise those features or regions not known. In normal image processing applications, this problem is somewhat less. Most features and regions can be accounted for. The limitations and techniques used by many authors are quite similar .

   Techniques used are widespread. Prototypes currently are primarily used, with future possibilities of automation, and operational applications. A substantial amount of work has been done on MR Images and the human brain, with most research currently focused on 3-D images.

13.4 CONCLUSIONS

An overview has to a lesser degree been provided for the KBS, Image Processing, and to a lesser degree Biomedical Applications. It is recognised that knowledge-based systems are now widely used. However, when combined with image processing, this is not the case. We have still a way to go before we have a number of applications that are based on this with a full move away from the experimental stage. Firstly, validation and verification issues need to be accounted for, so applications may be taken seriously. There is also a wide gap between the two fields of knowledge-based systems and image processing – a lack of communication. This may account for the current stage the applications or prototypes are in. Also, errors are more likely to occur as a result, for example, with Gonzales and Woods (1992 p. 8) diagram. It is also said that hybrid techniques are more useful for better control and representation (Todd-Pokropek, 1989) of images, which need to be incorporated into applications. However, since then recent years have seen a mixture of representation and control strategies, where often a top-down method is used for high-level processing and a bottom-up approach for low-levels.

We thus need to rely on past experience and have more communication between the two areas. This will allow for a better understanding of what is required and involved to improve and interpret images. If more authors from the two fields work together on the experiments and prototypes, then rapid advancements may be more likely to occur. This book represents a small step forward in this respect.

Other problems and limitations also need to be addressed, such as knowledge elicitation and acquisition. For interpretation purposes, a large quantity of information needs to be encoded into the system. Overcoming uncertainty is another. Currently, we may be in the situation where knowledge-based systems may generalise on information that they do not possess, resulting in erroneous conclusion. This could have undesirable consequences, its impact felt more so in the field of medicine. We presently have methods which use knowledge-based systems for image interpretation and understanding. However, we need to move to a stage where modularisation can be used to generalise on images, and not have interpretation specifically for that application.

Advances in automation and in the above method may be more likely to occur in image processing and knowledge-based systems generally than in the medical or biomedical field, due to unrecognised features or structures in medicine and biology. As Walker and Fox (1987, p. 262) conclude in their article:

> *'Biomedical images raise special challenges due to the variability and non-rigidity of biomedical structures. Ironically the very difficulty of interpreting these images may contribute to the development of general purpose vision systems because the simplifying assumptions which can make specialised applications relatively tractable cannot be adopted in the biomedical domain.'*

Besides these problems and limitations, a number of areas have been outlined by Walker and Fox (1987), that are issues for future research in biomedical imaging:

- exploring limitations in symbolic encoding techniques;
- identifying optimal information types to encode in a symbolic image description database;
- developing symbolic representations of spatial and temporal topologies characteristic biological objects and events;
- Using a non-image domain to constrain interpretation of image elements;
- deciding the degree low level image capture techniques ought to be controlled by knowledge interpretation processes.

Appendix A:

Exercises and Further Reading

D. GRAHAM, D. MARTLAND and A. BARRETT

A.1 EXERCISES

Some of these exercise deliberately require reading beyond the scope of this book. This is in recognition of the limited coverage possible. Several of the questions therefore aim to extend the knowledge and reading of readers.

A.1.1 Knowledge-Based Systems: An Overview

Introduction to Knowledge-Based Systems

Background to Knowledge-Based Systems

1. a) Discuss the following:
    'Artificial Intelligence is:
    the enterprise of simulating human intelligence;
    the enterprise of understanding human intelligence;
    an engineering discipline within computer science.'
    b) (i) Criticise Turing's criteria for computer software being 'intelligent'.
    (ii) Describe and justify your own criteria for computer software to be considered 'intelligent'.
    (iii) Create and justify your own definition of artificial intelligence.

What is a Knowledge-Based System?

2. a) What are the basic components of a knowledge-based system?
    b) How do knowledge-based systems differ, if at all, from expert systems?

    c) List some of the domains where knowledge-based systems have been applied.

Knowledge-Based System Architectures and Applications

3. a) Describe five broad categories under which software tools for building knowledge-based systems may be listed.

    b) Compare and contrast these categories. What are the strengths and weaknesses of each?

    c) Discuss some of the (current research) problems in building knowledge-based systems.

Knowledge Elicitation

4. a) Distinguish between knowledge acquisition and knowledge elicitation.

    b) Why is knowledge acquisition an indispensable part of expert system development?

    c) Describe a knowledge acquisition methodology with which you are familiar.

    d) Describe two knowledge acquisition systems with which you are familiar. Compare and contrast the two approaches employed.

Knowledge Representation

5. a) Make arguments for and against the use of formal logic in representing common-sense knowledge.

    b) Provide short accounts of the following mechanisms employed by semantic networks:
      (i) Inheritance;
      (ii) Demons;
      (iii) Defaults;
      (iv) Perspectives

    c) What are the powers and limitations of semantic networks? Discuss.

6. a) What is a frame system? What, if any, are its advantages compared to semantic networks? Illustrate with your own examples.

    b) How do objects compare with frames?

    c) Give evidence from your own experience that suggests a script-like or frame-like organisation of human memory.

    d) In the context of scripts, what are the script match and between-the-lines problems?

General

7. a) Give short accounts of following:
    (i)   Natural Language Understanding and Semantic Modelling
    (ii) Modelling Human Performance
    (iii) Languages and Environments for AI
    (iv) Vision
    (v)   Expert Systems
    (vi) Knowledge Elicitation

Neural Networks

These exercises are suggestions for further activity related to neural networks and image processing, and may require some programming effort. However, it may be possible to use simple tools, such as a spreadsheet package, or to use an existing program or toolkit (such as Matlab) in order to reduce the programming required. There are an increasing number of programs written in Java which are available via the World Wide Web – some of these can be used immediately without the need for modification. A list of suitable WWW sites is given at the end of these exercises.

1. Using the perceptron training algorithm, train a threshold logic unit to realise the following Boolean functions:

    x_1 OR x_1x_2
    x_1 AND x_2
    x_1 AND (NOT x_2)

    This exercise can be carried out with a spreadsheet. Remember that an additional bias input (x_3) will be required, set permanently to 1. The weights will then be $w_1, w_2, w_3,$. The columns of the spreadsheet should contain:

    $x_1, x_2, x_3, w_1, w_2, w_3,$ activation A, H(A),
    desired output $d(x_1, x_2, x_3)$, diff

    where $A = x_1.w_1 + x_2.w_2 + x_3.w_3$, and diff $= d(x_1 x_2 x_3) - H(A)$

    H(A) is the actual output of the unit obtained by applying the Heaviside function to A. The update rule for the weights is applied from one row to the next:

    $$w_i' = \begin{cases} w_i + xi, \text{if diff} = 1 \\ w_i - x_i, \text{if diff} = -1 \end{cases}$$

    which can be rewritten as $w_i' = w_i + \text{diff} \times x_i$, where i takes values from 1 to 3.

On each row, the values of x_1 and x_2 are chosen randomly from the set {0, 1}

This exercise can also be carried out by conventional programming.

2. Modify the spreadsheet of exercise 1 to perform training using the LMS algorithm. This requires the addition of a learning rate coefficient (lr), and a modification of the update rule:

$$w_i' = w_i + diff_2 \times lr \times x_i$$
$$\text{where } diff_2 = d(x_1, x_2, x_3) - A.$$

Monitor the mean squared error as this algorithm proceeds.

Note how this method does not converge directly to a solution, but gets gradually closer. Also, the output of the threshold logic unit is not used during training, but the difference between the desired output and the unit activation is minimised over the training set.

The final solution is obtained by choosing a suitable thresholding function for the resulting activation – typically H(A-0.5).

3. Consider a rectangular retinal grid, which contains values 0, 1. For example, consider the 5 by 5 grids:

00100	10001
00100	01010
11111	00100
00100	01010
00100	10001

which represent the characters + and x respectively.

Choose a set of characters to recognise on an appropriate grid size – for example the digits 0, 1, 2 . . . 9 or the letters a . . . z, and construct a suitable training set. Choosing a small grid size will speed up the processing, but will reduce the number of characters which can be represented satisfactorily.

a) Try to classify the characters using an appropriate number of threshold logic units – (eg 10 for digits, 26 for letters) – using either the perceptron training rule, or the LMS algorithm. Only one unit should respond with a 1 as output for any input pattern.

Is this feasible for the character representations you have chosen? Will this method work for any set of characters?

b) Now try to classify the characters using a multi-layer perceptron (try a single hidden layer with 10 units).

Can you classify all the characters satisfactorily with this method?

[For details of the back propagation algorithm, see Haykin – also the Matlab system includes a toolkit which you might find helpful for this task]

You can modify this exercise by applying noise to each pattern in the training set, and also by applying noise to patterns presented for recognition. This will be more representative of real world pattern classification problems.

When performing optimisation using these algorithms it is instructive to monitor the error (either in batches or for each pattern) as the algorithm proceeds. Poor results will be obtained with the LMS or back propagation algorithm if the learning rate is too large or too small – in the first case the algorithm may not converge to a solution, while in the second case convergence may be very slow.

4. Feedforward networks with a structure n-m-n (n inputs, m hidden units, and n outputs), where

$$m < n$$

can be used for some interesting problems. One possibility is to train such a network as an auto-associator, with the inputs being either real valued (grey levels) or 0, 1 (binary). The network can then be viewed as being in two parts – firstly the input to the hidden layer, secondly the hidden layer to the outputs – which can be considered as an encoder and a decoder respectively. If the outputs of the hidden units are quantised to a relatively few levels, the network can be treated as a simple data compression/decompression system. For example, a network with 64 – 16 – 64 structure can be used to encode and decode image data from an 8 by 8 grid. Suppose that the input data uses 8 bits per pixel for an acceptable image quality, and that the hidden layer is quantised to 4 levels, which thus requires 2 bits/pixel. The data compression ratio which would be achieved is thus approximately

$$\frac{64 \times 8}{16 \times 2} = 16 : 1$$

Note that this assumes that the storage required for the network weights is small compared with the storage requirements for the input image. Is this a valid assumption?

Train such a network with blocks of 8 × 8 data taken at random from a digital image, until the outputs are sufficiently similar to the inputs. Then process the whole image by scanning the network over the whole image. Evaluate the degradation in the image by visual inspection, and also by measuring the total mean squared error.

Modify the trained network by quantising the output of the hidden units to 8, 4 and 2 levels respectively (which will require 3, 2, and 1 bits), and again evaluate the resulting compression/decompression system by visually inspecting the output when presented with the trained image, and also by measuring the total mean squared error.

Further experiments can be made by evaluating this system on images which were not used for training.

For further information, see the book by Hassoun.

Compare this method of image data compression with the discrete cosine transformation based JPEG algorithm, and the fractal algorithm based on iterated function systems (see Barnsley (1992)).

A.1.2 Image Processing

1. A function $f(x)$ is evaluated for values of x corresponding to $x = 0, 1, 2, 3$; and has the values 2, 3, 4 and 4 respectively.
   a) Determine the Fourier coefficients $F(0)$ and $F(2)$ in the corresponding Fourier series:
   b) The following values are to be submitted as input to the Fast Fourier Transform:
      2, 0, 4, 0, 2, 1, 4, 3
   Determine the actual input values after reordering by the method of successive doubling. Confirm the results by the method of bit reversal.

2. A Fourier transform of the function $f(x)$ evaluated for values of x at 0, 1; yielded the transform pairs $F(0) = (1, 0)$ and $F(1) = (0, 1)$ respectively.
   a. Determine the value $f(1)$.
   b. Given the two sequences $f(x)$ and $g(x)$ defined for values of $x = 0, 1, \ldots, 31$;
   Derive an expression for the proportional time taken to determine the correlation of the two sequences using the FFT.

3. Given the set of gray levels $r(k)$ defined for $k = 0, 1, 2, 3$, and distributed as shown in the image below:
   a) obtain the equalised image and comment on the result.

1	1	2	3
0	1	3	2
3	2	0	1
3	2	2	0

b) It is decided to modify the image according to a set of density values p(k) defined for k = 0, 1, 2, 3, as 0.18, 0, 20, 0.25 and 0.10 respectively. Show the effect of applying this specified distribution to the image.

4. For the image below, determine the effect of sharpening processes on the pixel with a gray level value of 4.

a)

0	2	1	2
3	4	0	0
1	1	0	2
2	1	0	0

b) For the same image, determine the effect of smoothing processes applied to the same pixel.

5. For the following image:

a) use the appropriate template to confirm the existence of a point with a gray level of value 5

4	1	1	1
2	5	0	1
1	0	4	1
0	1	0	0

b) Confirm the existence of a line of slope $-45°$ in the above image.

c) Given the two points A and B whose corresponding co-ordinate pairs are (1, 0) and (1, 3) respectively; show the mapping of A & B under the parametric form of the Hough transform.

6. The following gray levels represent a line segmented from an image:

0, 3, 1, 5, 2, 2, 4, 1

Determine the smoothness of the line and its relative flatness.

A.2 FURTHER READING

A.2.1 Knowledge-Based Systems

There are now an abundance of texts on both Artificial Intelligence and Knowledge-Based or Expert Systems. A few Texts recommended and endorsed by my students now follow.

Artificial Intelligence

Still one of the most readable books is Winston (1992). Clear explanations of Artificial Intelligence topics, especially search strategies.

Knowledge-Based Systems

Probably the best book on knowledge-based systems by far is Jackson (1990). His treatment of knowledge-based system applications is particularly admirable. Applications discussed in the context of knowledge acquisition, consequently links between systems often in different domains, are shown.

Other well-written texts include Luger and Stubblefied (1992), and Harmon and King (1976). Although the latter is comparatively old now, basic concepts are explained in layman's terms.

Additionally, readers should refer to the extensive list of references given in the next section.

Neural Networks

This chapter has only covered the subject of neural networks very briefly. There has been a great deal of interest in this area during the last 10 years, and many papers and books have been written about it.

Here are a few ideas for further reading:

An Introduction to Neural Computing, Igor Aleksander and Helen Morton, Chapman and Hall, London, 1990.

This book is relatively easy to read, and is one of the few texts to include sections on Aleksander's own networks, based on the use of random access memory units as neural units. It also covers several other artificial neural network types.

Perceptrons: an introduction to computational geometry, Marvin Minsky and Seymour Papert, MIT Press, 1969

This book is now a classic. It is still worth reading as it gives detailed descriptions of the work which led Minsky and Papert to reject the use of simple perceptrons as useful learning systems. Now we know that it is possible to use more complex models successfully, but the basic arguments are still valid.

Neural Computing: theory and practice, P. D. Wasserman, Van Nostrand Reinhold, New York, 1989.

This book is approachable, and gives a good basic overview of the subject.

Neural Networks: a comprehensive foundation, Simon Haykin, Macmillan, 1994.

This book is a very good, comprehensive text, but it does contain a fair amount of mathematics. It is intended as a graduate level text.

Artificial Neural Networks, Mohamad H. Hassoun, MIT Press, 1995.

This book is excellent, and very detailed. It is also highly mathematical, and appears to be aimed largely at graduate students.

Neural Networks: a systematic approach, R. Rojas, Springer-Verlag, Berlin, 1996.

This book offers a good overview of much of the field.

Neurocomputing: foundations of research, James A. Anderson and Edward Rosenfeld (Eds), MIT Press, Cambridge, Mass, 1988.

This book collects together many interesting and important papers from over 50 years of neural network research.

Neurocomputing 2: directions of research, James A. Anderson, Edward Rosenfeld and Andras Pellionisz (Eds), MIT Press, Cambridge, Mass, 1990.

This is the follow up to the earlier collection.

The Molecular Biology of the Cell, Bruce Alberts, Dennis Bray, Julian Lewis, Martin Raff, Keith Roberts, James D. Watson. Garland, (3rd edition), New York, 1994

This is a detailed book about cell biology, so much of it is not directly relevant. However, the sections on neurons, and the operation of neurotransmitters are fascinating, and give considerable insight into the way biological nervous systems function, and to the background of artificial neural systems.

IEEE Computer, Special Issue on Neural Networks, March 1988.

This special issue contains several interesting articles.

Neural Networks, 1(1), 1988.

There are several overview articles in the very first issue of the Neural Networks journal.

Journals to read regularly include Neural Networks, and the IEEE Transactions on Neural Networks.

A.2.2 Image Processing

Further reading:

Ballard, D. H. (1981). 'Generalizing the Hough Transform to Detect Arbitrary Shapes.' Pattern Recognition, vol. 13, no. 2, pp 111–122.

Bouman, C., and Liu, B. (1991). 'Multiple Resolution Segmentation of Textured Images.' IEEE Trans. Pattern Recognition Anal. Machine Intell., vol. 13. no. 2, pp 99–113.

Brummer, M. E. (1991). 'Hough Transform Detectionof the Longitudinal Fissure in tomographic Head Images.' IEEE Trans. Biomed. Images, vol. 10, no. 1, pp 74–83.

Chen, D., and Wang, L. (1991). 'Texture Features Based on Texture Spectrum.' Pattern Recog., vol. 24, no. 5, pp 391–399.

Davis, L. S. (1982). 'Hierarchical Generalized Hough Transforms and Line-Segment Based Generalised Hough Transforms.' Pattern Recog., vol. 15, no. 4, pp 277–285.

Gupta, L., Mohammad, R. S., and Tammana, R. 1990. 'A Neural Network Approach to Robust Shape Classification.' Pattern Recog., vol. 23, no. 6, pp 563–568.

Gonzalez, R. C. and Woods, R. E. (1992). *Digital Image Processing.* Addison-Wesley Publishing Company.

Haddon, J. F., and Boyce, J. F. 1990. 'Image Segmentation by Unifying Region and Boundary Information.' IEEE Trans. Pattern. Machine Intell., vol. 12, no. 10, pp 929–948.

Perez, A., and Gonzalez, R. C. (1987). 'An Iterative Thresholding Algorithm for Image Segmentation.' IEEE Trans. Pattern Anal Machine Intell., vol. PAMI-9, no. 6, pp 742–751.

Shih, F. Y. C., and Mithcell, O. R. (1989). 'Treshold Decomposition of Gray-Scale Morphology into Binary Morphology.' IEEE Trans. Pattern Anal. Machine Intell., vol. 11, no. 1, pp 31–42.

References

Aikins, J. S. (1983). Prototypical knowledge for expert systems. *Artificial Intelligence*, **20**, 163–210.

Alberts, B., Bray, D., Lewis, J., Raff, M., Roberts, K. and Watson, J. D. (1983), The Molecular Biology of the Cell, Garland, New York.

Aleksander, I. and Morton, H. (1990), An Introduction to Neural Computing, Chapman and Hall, London.

Aleksander, I., Stonham, J. and Wilkie, B. A. (1983), Computer vision for robots, some comparisons, in 'Artificial vision for robots', Aleksander, I. (ed), Kogan Page, London.

Alemany, J. and Kasturi, R. (1988). A Computer Vision System for Interpretation Paper-based Maps. *SPIE Applications of Digital Image Proceedings X*. San Diego **829**. In: Gilles, A. C. and Hart, A. (1988).

Alexander, J. H., Freilling, M. J., Shulman, S. J., Staley, J. L., Rehfuss, S. and Messick, S. L. (1986). Knowledge level engineering: ontological analysis. *Proceedings of the National Conference on Artificial Intelligence*, pp 936–8.

Alvey, P. (1983). Problems of designing a medical expert system. In: Jackson, P. (1990).

Alvey, P. L., Greaves, M. F. (1987). Observations on the development of a high performance system for leukaemia diagnosis. In: *Research and development in expert systems III*, Bramer, M. A. (Ed.), Cambridge University Press, pp 99–110.

Anderson, J. A. and Rosenfeld, E. (Eds) (1990), Neurocomputing: foundations of research, MIT Press, Cambridge, Mass.

Arbab, B., Michie, D. (1985). Generating expert rules from examples in Prolog, *TIRM-85-010*, Turing Institute, Glasgow.

Aryanpur, S. (1988). Expert Systems: A buyer's market? *Systems International*, April 1988, pp 29–32.

Baba, N. (1989), A new approach for finding the global minimum of error function of neural networks, Neural Networks, **2**, 367–373.

Bachant, J., McDermott, J. (1984). R1 Revisited: Four Years in the Trenches. *A.I. Magazine* **5**(3).

Banathy, B. H. (1988). Matching design methods to system type. *Systems Research* **5**(1), pp 27–34.

Banerjee, A., Majunder, A. K., Basu, A. (1994). A Knowledge-based system for using multiple expert modules for monitoring leprosy – an endemic disease. *IEEE Transactions on Systems, Man, Cybernetics* **24**(2), pp 173–186.

Barnsley, M and Hurd, L.P. (1993). Fractal image compression, A.K. Peters.

Barr, A., Feigenbaum, E. A., eds. (1981). *The Handbook of Artificial Intelligence* Vol 1. Los Altos CA: Morgan Kaufmann.

Basden, A. (1984). On the application of expert systems. In: *Developments in expert systems*, Coombs, M. J. (Ed.), London: Academic Press, pp 59–75.

Basra, J.K (1995). A Review of Knowledge-Based Biomedical Image Processing. Unpublished project submitted in partial fulfilment of the requirements for the degree of Bachelor of Science. Department of Computer Science and Information Systems, Brunel University, June, 1995.

Berry, D. C., Broadbent, D. E. (1986). Expert systems and the man-machine interface. *Expert Systems* **3**(4), pp 228–231.

Bimbo, A. D., Campani M., and Nesi P. (1992). Using 3D Spatial relations for image retrieval by Contents. *Technical Report*, Italy. In: Chang, S. and Hsu, A. (1992).

Blackler, T. (1989). Articulating Expertise: The Use of Competence Models in Knowledge Elicitation. Unpublished project submitted in partial fulfilment of the requirements for the degree of Bachelor of Science, Department of Computer Science, Brunel University, June, 1989.

Bledsoe, W. W. and Browning, I. (1959), Pattern Recognition and reading by machine, Proceedings of the Eastern Joint Computer Conference, 225–232.

Bobrow, D. G., Sefrik, M. (1983). *The LOOPS Manual Xerox Corporation*. In: Jackson P. (1990).

Bobrow, D. G., Winograd, T. (1977). An overview of KRL, a knowledge representation language. *Cognitive Science* **1**, pp 3–46.

Boose, J. H. (1986). *Expertise Transfer for Expert System Design*. New York: Elsevier.

Brachman, R. J. (1985a). I lied about the trees. *AI Magazine* 6(3).

Brachman, R. J., Fikes, R. E., Levesque, H. J. (1983). KRYPTON: A functional approach to knowledge representation. *IEEE Computer*, September, 1983.

Breuker, J. A., Wielinga, B. J. (1983). Analysis techniques for knowledge based systems, Part 2: methods for knowledge acquisition. Report 1.2, Esprit Project 12 (Memorandum 13 of the Research Project: The Acquisition of Expertise). December 1983, Amsterdam: University of Amsterdam.

Breuker, J. A., Wielinga, B. J. (1987a). Use of models in the interpretation of verbal data. In: *Knowledge acquisition for expert systems: a practical handbook*, Kidd, A. L. (ed.), New York, NY: Plenum, pp 17–44.

Breuker, J. A.; et al. (1987). *Model-driven knowledge acquisition: interpretation models*. Deliverable task A1, Esprit Project 1098, Amsterdam, University of Amsterdam.

Brigham, E. O. (1974). *The Fast Fourier Transform*, Prentice-Hall, Englewood Cliffs, N. J.

Brookes, R., Greiner, R. and Binford, I. (1979). The ACRONYM model-based vision system. *Proceedings ISCAI*. Tokyo **6**, pp 105–113. In: Walker N. and Fox J. 1987.

Browse (1982). Knowledge-based Interpretation Using Declarative Schema. Vancouver, *Technical Report* TN. In: Rao, A. R. and Jain, R. (1988).

Buchanan, B. G. and Feigenbaum, E. A. (1978). DENDRAL and META-DENDRAL: their applications dimension. *Artificial Intelligence*, **11**, 5–24.

Buchanan, B. G., Barstow, D., Betchtel, R., Bennet, J., Clancey, W., Kulikowski, C., Mitchell, T. M. and Waterman, D. A. (1983). Constructing an expert system. In: Hayes-Roth *et al.*, pp 127–167.

Buchanan, B. G., Duda, R. O. (1983). Principles of rule-based expert systems. *Advances in Computers* **22**, pp 163–216.

Carnegie Group (1985). *Knowledge Craft 3.0 Reference Manual*. Pittsburgh PA: Carnegie Group.

Carpenter, G. A, Grossberg, S. and Reynolds, J. H. (1991), ARTMAP: Supervised real-time learning and classification of nonstationary data by a self-organizing network, Neural Networks, **4**, 565–580.

Carpenter, G. A, Grossberg, S. and Rosen, D. B.(1991), Fuzzy ART: SFast stable learning and categorization of analog patterns by an adaptive resonance system. Neural Networks, **4**, 759–771.

Chandrasekaran, B. (1983). Towards a taxonomy of problem solving types. *AI Magazine*. Winter/Spring, pp 9–17.

Chandrasekaran, B. (1986). Generic tasks in knowledge-based reasoning: high level building blocks for expert systems design. *IEEE Expert* **1(3)**, pp 23–30.

Chandrasekaran, B. (1987). Towards a functional architecture for intelligence based on generic information processing tasks. In: *Proceedings of the Tenth International Conference on Artificial Intelligence*. Milan, Italy, August 1987, pp 1183–1192.

Chandrasekaran, B. (1989). Generic tasks as building blocks for knowledge-based systems:the diagnosis and routine design examples. *The Knowledge Engineering Review*, Vol 3, Part 3, September 1989, pp 183–210.

Chang, S. and Hsu, A. (1992). Image Information Systems: Where do we go from here? *IEEE Transactions on Knowledge and Data Engineering* **4**(5), pp 431–441.

Clancey, W. J. (1983). The epistemology of a rule-based expert system – a framework for explanation. *Artificial Intelligence* **20**, pp 215–251.

Clancey, W. J. (1985). Heuristic Classification. *Artificial Intelligence* **27**, pp 289–350.

Clancey, W. J. (1989). Viewing Knowledge Bases as Qualitative Models, *IEEE Expert*, Summer 1989, pp 9–23.

Clancey, W. J. and Letsinger, R. (1981). NEOMYCIN: reconfiguring a rule-based expert system for application to teaching. *Proceedings of the Seventh International Joint Conference on Artificial Intelligence (IJCAI-81)*. Vancouver, Canadar, 24–28 August, 1981, pp 829–836.

Clark, D. A., Rawlings, C. J., Barton, G. J. and Archer, I. (1990). Genetic Map construction with constraints. *Proceedings of 2nd International Conference on Intelligent Systems for Molecular Biology*, AAAI Press. In: Fox, J. and Rawlings, C. J. (1992).

Cooley, J. W. and Tukey, J. W. (1967). *An Algorithm for the Machine Calculation of Complex Fourier Series*. Math. of Comput., vol 19, pp 297–301.

Coombs, M. A. and Atly, J. L. (1984). Expert systems: an alternative paradigm. In: *Developments in expert systems*. Coombs, M. J. (ed.), London: Academic Press.

Cordingly, E. S. (1989). Knowledge elicitation techniques for knowledge-based systems. In: *Knowledge Elicitation: principles, techniques and applications*. Diaper, D. (ed.), Ellis Horwood, pp 89–175.

Davis, R. (1980b). Applications of meta knowledge to the construction, maintenance and use of large knowledge bases. In: Davis and Lenat (1980), pp 229–490.

Davis, R. and King, J. (1977). An overview of production systems. In: Elcock and Michie (1977), pp 300–32.

Dayhoff, J. (1990), Neural network architectures – an introduction, Con Nostrand Reinhold

DeKleer, J. (1986). An assumption-based TMS. *Artificial Intelligence*, **28**, 127–62.

Diaper, D. (1987a). Designing Systems for People: Beyond User Centred Design. In: *Software Engineering, Proceedings SEAS 1987 Conference.*

Diaper, D. (ed.) (1989). *Knowledge elicitation: principles, techniques and applications.* Chichester: Ellis Horwood.

Dowling, J. E. (1992). Neurons and networks: an introduction to neuroscience, Belknap Press of Harvard University Press, Cambridge, MA.

Doyle, J. (1979). A truth maintenance system. *Artificial Intelligence*, **28**, 231–72.

Edmunds, R. A. (1988). *The Prentice-Hall Guide to Expert Systems.* Englewook Cliffs, Prentice-Hall.

Eisner, J., Tedesco, L. and Villo, R. (1993). DentLE the Dental learning environment, A Prototype. In: *Sixteenth annual Symposium on Computer Applications in Medical Care.* Frisse, M. E. (ed.) (1993). New York, McGraw-Hill, pp 697–701.

Erman, L. D., Lsrk, J. S. and Hayes-Roth, F. (1986). Engineering intelligent systems: progress report on ABE. *Proceedings of Expert Systems Workshop*, Defense Advanced Research Projects Agency.

Evans, J. St B. T. (1987). Human biases and computer decision-making: a discussion of Jacob et al. *Behaviour and Information Technology* **6(4)**, pp 483–487.

Feigenbaum, E. A. (1977). The art of artificial intelligence: themes and case studies of knowledge engineering. *Proceedings of the 5th International Joint Conference on Artificial Intelligence*, pp 1014–29.

Feigenbaum, E. A. (1983). In: Feigenbaum, E. and McCorduck, P. (1983) *The Fifth Generation*, Addison-Wesley Pub. Co., pp 76–77.

Findler, N. V., (ed.) (1979). *Associative Networks.* New York: Academic Press.

Firebaugh, M. V. (1989). *Artificial Intelligence A Knowledge-based Approach.* Boston, PWS-Kent Publishing Co.

Forgy, C. L. (1982). Rete: a fast algorithm for the many pattern/many object pattern match problem. *Artificial Intelligence*, **19**, 17–37.

Fox, J. and Rawlings, C. J. (1992). Artificial Intelligence and Knowledge-based Systems in Molecular Biology. *The Knowledge Engineering Review* **9**(3) pp 287–300.

Fukushima, K. (1980), Neocognitron: a self organizing neural network moddel for a mechanism of pattern recognition unafffected by shift in position, Biological Cybernetics, **36**, 193–202.

Fukushima, K. (1988), Neocognitron: a hierarchical neural network capable of visual pattern recognition, Neural Networks, I., 119–130.

Gaines, B. R. (1986). The British Computer Society Expert Systems1986 Conference, state-of-the-art report from the AAAI Workshop on Knowledge Acquisition held on November 1986.

Gaines, B. R. (1989). Knowledge Acquisition and Technology. In: *Knowledge Acquisition for KBS*, Gaines, B. and Boose, J. (eds.), London: Academic Press.

Gale, W. A. (1986). *Artificial Intelligence and Statistics.* Reading MA: Addison-Wesley.

Gammack, J. G. and Young, R. M. (1985). Psychological techniques for eliciting expert knowledge. In: *Knowledge search acquisition for expert systems.* Bramer, M. A. (ed.), Cambridge: Cambridge University Press, pp 105–112.

Garcia, O. N. and Tzuu-Chien, Y. (1991). *Knowledge-based System: Fundamentals and Tools.* Los Angeles, IEEE Computer Society Press.

Geman, S. and Geman, D. (1984). Stochastic relaxation, Gibb's distribution, and the Bayesian restoration of images, IEEE. Transactions on Pattern Analysis and Machine Intelligence, **6**, 721–741.

Gilles, A. C. and Hart, A. (1988). Using KBS Ideas in Image Processing – A Case Study in Human-Computer Interaction. In: *Research and Development V Proceedings of Expert Systems '88 Eigth annual Technical Conference of BCS.* Kelly, B. and Rector, A. (eds.). Cambridge University Press, pp 257–269.

Gilliot, J. M., Amat, J. L. and Stamon, G. (1993). Thin-linear network extraction NEXSYS: a knowledge-based system for SPOT images. *Proceedings of the SPIE* **1771**, pp 88–98.

Goldberg, D. E. (1989), Genetic Algorithms in search, optimization and machine learning, Addison-Wesley, Reading, Mass.

Gonzales, R. C. and Woods, R. E. (1992). *Digital Image Processing.* Reading, Addison Wesley Publishing Co.

Gonzalez, R. C. and Wintz, P. (1977). *Digital Image Processing*, Addison-Wesley Publishing Company.

Goodall, A. (1985). *The guide to expert systems.* Oxford Learned Information.

Goodenough, D. G., Charlebois, D., Matwin, S. and Robson, M. (1994). Automating reuse of software for expert system analysis of remote sensing data. *IEEE Transactions on Geoscience and Remote Sensing* **32** (3) pp 525–533.

Graham, D. (1990). *Knowledge Elicitation: A Case Study in Computer Fault Diagnosis and Repair.* Department of Computer Science, Unpublished PhD Thesis, Brunel University, London.

Green, P. (1984). *MSc. Course Notes: Artificial Intelligence.* North Staffordshire Polytechnic.

Grimm, F. and Burke, H. (1993). An expert system for the selection and application of image processing subroutines. *Expert Systems* **10** (2) pp 61–74.

Grover, M. D. (1983). A pragmatic knowledge acquisition methodology. *Proceedings of the Eigth International Joint Conference on Artificial Intelligence (IJCAI-83).* Karlsruhe, West Germany, 8–12 August, 1983, pp 436–438.

Gupta, R. and Horowitz, E. (1991). *Object-oriented Database with Applications to CASE: Networks and VLSI CAD Englewood Cliffs.* In: Chang, S. and Hsu, A. (1992).

Hammond, P., Davenport, J. C. and Fitzpatrick, F. J. (1993). Logic-based integrity constraints and the design of dental prostheses. *Artificial Intelligence in Medicine* 5, pp 431–46, Elsevier.

Harmon, P. and King, D. (1985). *EXPERT SYSTEMS Artificial Intelligence in Business.* John Wiley & Sons, inc.

Hartline, H. R., Ratliff, F., and Miller, W. H. (1961) Inhibitory interactions in the retina and its significance in vision, in E. Florey (Ed), Nervous Inhibition, Pergamon Press, Oxford

Hassoun, M., H. (1995), Fundamentals of Artificial Neural Networks, MIT Press, Cambridge, Mass.

Hawkins, D. (1983). An analysis of expert thinking. *International Journal of Man-Machine Studies* 18(1), pp 1–47.

Hayes, P. J. (1974). Some problems and non-problems in representation theory. *Proc. AISB Summer Conference* 63–69. University of Sussex.

Hayes-Roth, B., Garvey, A., Johnson, M. V. and Hewett, H. (1987). *A Modular and Layered Environment for Reasoning About Action.* Technical Report No. KSL 86–38, Knowledge Systems Laboratory, Stanford University.

Hayes-Roth, F., Waterman, D. A. and Lenat, D. B. (eds.) (1983). *Building expert systems.* Reading, MA: Addison-Wesley.

Haykin, S. S. (1994). Neural Networks: a comprehensive foundation, Macmillan, New York.

Hebb, D. (1949), The Organization of Behaviour, Wiley, New York.

Hodgkin, A. L. and Huxley, A. F. (1952), A quantitative description of membrane current and its application to conduction and excitation in nerve, J. Physiol. **117**, 500–544.

Hopfield, J. J. (1982), Neural networks and physical systems with emergent collective computational abilities, Proceedings of the National Academy of Sciences, USA, **79**, 2445–2558

Hough, P. V. C. (1962). *Methods and means for Recognizing Complex Patterns.* U. S. Patent 3,069,654.

HSMO (1991). *Knowledge-Based Systems.* London Crown Copyright.

Intellicorp (1984). *The Knowledge Engineering Environment.* Mountain View CA: Intellicorp.

Jackson, A. (1986b). A little learning is a dang'rous thing. *Datalink.* 3 February, 1986, p 10.

Jackson, P. (1990). *Introduction to Expert Systems* (2nd edition). Reading, Addison-Wesley Publishing Company.

Jacobs, R. A. (1988), Increased rates of convergence through learning with adaptation, Neural Networks, 1(4), 295–230.

Johnson, L. and Johnson, N. E. (1987a). *Research Methods in Building Knowledge Based Systems.* Brunel University, May 1987.

Johnson, L. and Johnson, N. E. (1987b). Knowledge elicitation involving teachback interviewing. In: *Knowledge acquisition for expert system: a practical handbook.* Kidd, A. L. (ed.), New York, NY: Plenum, pp 91–108.

Johnson, N. E. (1990). *pers comm.* July 1990.

Johnson, P., Diaper, D. and Long, J. (1984). Tasks, skills and knowledge: task analysis for knowledge based descriptions'. In: *Human-Computer Interaction- INTERACT'84.* Shackel, B. (ed.), Elsevier Science Publishers B. V. (North-Holland), pp 499–503.

Johnson, P., Johnson, H. and Russel, F. (1988). *Collecting and generalising knowledge descriptions from task analysis data.* QMC, London.

Kauffman, S. A.

Keene, S. E. (1989). *Object-Oriented Programming in COMMON LISP.* Reading MA: Addison-Wesley.

Keravnou, E. and Johnson, L. (1986). *Competent Expert Systems*. Kogan Page.

Kidd, A. L. (ed.) (1987). *Knowledge acquisition for expert systems: a practical handbook*. New York, NY: Plenum Press.

Koch, C. and Segev, I. (1989), Methods in neuronal modeling: from synapses to networks, MIT Press, Cambridge, Mass.

Kohonen, T. (1977), Associative memory – a system theoretical approach, Springer-Verlag, Heidelberg.

Kohonen, T. (1984), Self Organization and Associative Memory, Springer-Verlag, Berlin.

Kohonen, T. (1995), Self-organizing maps, Springer-Verlag, Heidelberg.

Kosko, B. (1988), Bidirectional associative memories, IEEE Transactions on Systems, Man and Cybernetics, SMC-18, 49–60.

Kosko, B. (1992), Neural networks and fuzzy systems, Prentice Hall, Englewood Cliffs, New Jersey.

Kowalski, R. A. (1979). *Logic for problem solving*. Amsterdam: North-Holland.

Krause and Clark (1993). In: Palmer, J. (1995).

Kujawinska, M. (1993). Expert system for analysis of complicated fring benefits. *Proceedings of the SPIE – The International Society for Optical Engineering* 1755 pp 252–257.

Kunz, J. C., Kehler, T. P. and Williams, M. D. (1984). Applications development using a hybrid AI development system. *AI Magazine*, 5(3).

Lehky, S. R. and Sejnowski, T. J. (1988), Network model of shape-from-shading: neural function raises from the receptive and prjective fields, Nature, **333**, 452–454.

Little, W. A. (1974), The existence of persistent states in the brain, Mathematical Bioscience, **19**, 101–120.

Little, W. A., Shaw, G. L. (1974), A statistical theory of short and long term memory, Behav. Bol. **14**, 115.

Lowe, G. (1987). *AI*, **31** pp 374–378. In: Tjahjadi, T. and Henson, R. (1989).

Luger, G. F. and Stubblefield, W. A. (1993). *Artificial Intelligence: structures and strategies for complex problem solving (2nd edition)*. Benjamin-Cummings Pub. Co. Inc.

Marr, D. (1982). *Vision*. W. H. Freeman and Company, San Francisco, California.

Martil, R. A. (1989). The Role of Machine Induction in Knowledge Acquisition. Unpublished MSc. dissertation in Intelligent Systems. Department of Electrical Engineering, Brunel University, April, 1989.

Masterman, M. (1961). Semantic message detection for machine translation, using Interlingua. *Proc. of the 1961 International Conference on Machine Translation*.

Matsuyama, T. (1992). Expert Vision. *Journal of the Institute of Television Engineers of Japan* **46**(11), pp 1410–1418.

McAllester, D. (1980). An Outlook on Truth Maintenance. Report No. AIM-551, Artificial Intelligence Laboratory, Massachusetts Institute of Technology.

McCullogh and Pitts, W. (1943), A logical calculus of ideas immanent in nervous activity, Bulletin of Mathematical Biophysics, **5**, 115–133.

McDermott, J. (1981). R1, The formative years. *AI Magazine* (Summer).

Mead, C. (1989), Analog VLSI and neural systems, Addison-Wesley, Reading, Mass.

Meyers, C. D., Fox, J., Pegram, S. M. and Greaves, M. F. (1993). Knowledge acquisition for expert systems: experience using EMYCIN for leukaemia diagnosis. In: Jackson, P. (1990).

Michalski, R. S. (1983). A Theory and Methodology of Inductive Learning. *Artificial Intelligence*, **Vol. 20**, pp 111–161.

Michie, D. (1983). Inductive Rule Generation in the Context of the Fifth Generation. *Proceedings of the International Machine Learning Workshop*.

Minsky, M. (1975). A framework for representing knowledge. In Brachman and Levesque (1985).

Minsky, M. and Papert, S. (1969), Perceptrons: An introduction to computational geometry, MIT Press, Cambridge, Mass.

Musen, M. A., Fagan, L., Coombs, D. M. and Shortliffe, E. H. (1987). Use of a domain model to drive an interactive knowledge-editing tool. *International Journal of Man-Machine Studies*, **26**, 105–21.

Mylopolous, J. and Levesque, H. J. (1984). An overview of knowledge representation. In: Brodie *et al.* (1984).

Naeym-Rad, F., de Souza Almeida, F. and Trace, D. (1993). IMR Entry. In: *Sixteenth anual Symposium on Computer Applications in Medical Care*. Frisse, M. E. (ed). New York, McGraw-Hill, pp 783–784.

Neale, I. M. (1988). First generation expert systems: a review of knowledge acquisition methodologies. In: *The Knowledge Engineering Review*, **Vol. 3, No. 2**. Cambridge University Press, June, 1988, pp 103–143.

Newell, A. (1982). The knowledge level. *Artificial Intelligence* **18**, pp 87–127.

Newell, A. and Simon, H.A. (1972). Human problem Solving. Englewood Cliffs NJ: Prentice-Hall.

Niblack, W. (1986). *An Introduction to Digital Image Processing.* 2nd edn, Prentie Hall, Englewood Cliffs, N. J.

Palmer, J. (1995). The Application of Knowledge Based Systems to Mechanical Computer Aided Design. PhD Interim Report. Department of Computer Science and Information Systems, Brunel University, London.

Parker, D. B. (1985). Learning Logic, TR-47, Center for Computational Research in Economics and Management Science, MIT, (April 1985).

Patterson, D. W. (1990). *Introduction to Artificial Intelligence and Expert Systems.* Englewoods, Prentice Hall.

Perkins, W. J., Hashmi, S. and Jordan, M. (1993). 3D computer modelling system for the study of biological structures. In: *Medical and Biomedical Engineering and Computing.* November 1993. Pople *et al.* (1974?).

Plant, T., Scrivener, S. A. R., Schappo, A. and Woodcock, A. (1989). Usage and generality of knowledge in the interpretation of diagrams. *Knowledge-based Systems*, **2** (2) pp 99–108.

Pople, H. E. (1977). The composite hypotesis in diagnostic problem solving: an exercise in synthetic reasoning. Proceedings of the 5th International Joint Conference on Artificial Intelligence, pp. 1030–7.

Post, E. L. (1943). Formal reductions of the general combinatorial decision problem. *American Journal of Mathematics*, **65**, 197–268.

Prabhakaran, N., Subbarao, W. and Penagos, J. (1990). An expert system approach for optical image processing and anlysis. In: *Proceedings of the IASTED International Symposium. Expert systems Theory and applications.* Hazma, M. H. (ed.). Calgary, Acta Press, pp 5–6.

Rao, A. R. and Jain, R. (1988). Knowledge Representation and Control in Computer Vision Systems. *IEEE Expert Systems and their Applications*, pp 65–79.

Rojas, R. (1996), Neural Networks: a systematic approach, Springer-Verlag, Berlin.

Rosenblatt, F. (1962), Principles of Neurodynamics, Spartan Press, Washington.

Roth, J. P. (1990). *Case Studies of Optical Storage Applications.* Meckler Publishing.

Rumelhart, D. E., Hinton, G. E., and Williams, R. J. (1986), Learning internal representations by error propagation, in Parallel Distributed Processing: Explorations in the Microstructure of Cognition, Vol. 1, D. E. Rumelhart, J. L. McClelland and the PDP Research Group(eds), MIT Press, Cambridge, Mass

Samadani, R. (1993). Introduction to Digital Image Processing and Computer Vision. In: *Advances in Computer Methods for Systematic Biology AI, Database, Computer Vision.* Baltimore, R. F., (ed.). London, John Hopkins University Press, pp 353–362.

Schank, R. C. (1975). *Conceptual Information Processing.* Amsterdam: North-Holland.

Schank, R. C. and Abelson, R. (1977). *Scripts, Plans, Goals and Understanding: Five Programs Plus Minatures.* Hillside, N. J.: Erlbaum.

Schank, R. C. and Colby, K. M (eds) (1973). *Computer Models of Thought and Language.* San Francisco: Freeman.

Schank, R. C. and Nash-Webber, B. L. (eds) (1975). *Theoretical Issues in Natural Language Processing.* Association for Computational Linguistics.

Schank, R. C. and Reiger, C. J. (1974). Inference and the computer understanding of natural language. *Artificial Intelligence* 5(4): 373–412.

Shapira, R. (1984). *IEEE Trans. PAMI*-6, pp 789–794. In: Tjahjadi, T. and Henson, R. (1989).

Shaw, M. L. G. and Gaines, B. R. (1987). KITTEN: Knowledge initiation and transfer tools experts and novices. *International Journal of Man-Machine Studies* **27**, pp 251–280.

Shortliffe, E. H. (1976). *MYCIN: Computer-Based Medical Consultation.* American Elsevier, New York.

Shortliffe, E. H., Scott, A. C., Bischoff, M. B., van Melle, W. and Jacobs, C. D. (1981). ONCOCIN: an expert system for oncology protocol management. *Proceedings of the 7th International Joint Conference on Artificial Intelligence*, pp 876–81.

Shu, S. and Freeman, H. (1988). An Expert system for Image Separation. *SPIE Applications of Digital Image Processing* **829** pp 240–252.

Solis, J. and J-B Wets, R. (1981), Minimization by random search techniques, Mathematics of Operations Research, **6**, 19–30

Sowa, J. F. (1984). *Conceptual Structures: Information Processing in Mind and Machine.* Reading, MA: Addison-Wesley.

Stevens, C. F. (1966) Neurophysiology: a primer, Wiley, New York.

Tamura *et al.* (1983). In Basra, J.K. (1995).

Tjahjadi, T. and Henson, R. (1989). A Knowledge-based System for Image Understanding. *Third International Conference on Image Processing and its Applications*, pp 82–92.

Todd-Pokropek, A. (1989). Expert Systems for handling medical images. *Processing of the SPIE: Science and Engineering of Medical Imaging* **1137**, pp 152–161.

Touretzky, D. S. (1986). *The Mathematics of Inheritance Systems*. Los Altos, CA: Morgan Kaufmann.

Turner, R. (1984). Logics for Artificial Intelligence. Chichester: Ellis Horwood Ltd.

van Melle, W. S. (1981). System Aids in Constructing Consultation Programs. Ann Arbor MI, UMI Research Press. In: Jackson, P. (1990).

vanMelle, W.S (1979). A domain-independent production-rule system for consultation programs. *Proceedings of the IJCAI-79*, pp 923–925.

Walker, N. and Fox, J. (1987). Knowledge-based interpretation of images: a biomedical perspective. *The Knowledge Engineering Review* **2** (4), pp 249–263.

Wasserman, P. D. (1989), Neural Computing: theory and practice, Van Nostrand Reinhold, New York

Waterman, D. A. (1986). *A guide to expert systems*. Reading, MA: Addison-Wesley.

Werbos, P. (1974), Beyond Regression: New tools for prediction and analysis in the behavioral sciences, PhD dissertation, Harvard University, Cambridge, Mass.

Widrow, B. and Hoff, M. E., Jr (1960), Adaptive Switching Circuits, IRE Western Electric Show and Convention Record, part 4, 96–104

Wielinga, B. J. and Breuker, J. A. (1985). Interpretation of verbal data for knowledge acquisition. In: *Advances in artificial intelligence*. O'Shea, T. (ed.). Amsterdam: North-Holland, pp 3–12.

Wilks, Y. A. (1972). *Grammar, Meaning and the Machine Analysis of Language*. London: Routledge & Kegan Paul.

Wilson, M. (1989). Task models for knowledge elicitation. In: *Knowledge Elicitation: principles, techniques and applications*. Diaper, D. (ed.). Chichester: Ellis Horwood, pp 197–219.

Wilson, M. D., Barnard, P. J. and MacLean, A. (1986). Task analyses in human-computer iteraction. *IBM Hursley Human Factors Report HF122*.

Winstan, P. H. (1992). Artificial Intelligence, 3rd edition. Addisan-Wesley.

Woods, W. (1975), What's in a link: foundations for semantic networks.

Young, R. M. (1988). Role of intermediate representations in knowledge elicitation. In: *Research and development in expert systems IV*. Moralee, D. S. (ed.). Cambridge: Cambridge University Press, pp 287–288.

Zahzah, E., Desachy, J. and Zehna, M. (1992). A fuzzy connectionist knowledge-based image interpretation system. In: *ICIP Proceedings of the 2nd Singapore International Conference on Image Processing*. Srinivasa, V., Heng, O. S. and Hock, A. W. (eds.). Singapore, World Scientifc, pp 404–408.

Author Index

Subject Index